How Should I Know?

NEW DIRECTIONS IN MATHEMATICS AND SCIENCE EDUCATION
Volume 8

Scope

Mathematics and science education are in a state of change. Received models of teaching, curriculum, and researching in the two fields are adopting and developing new ways of thinking about how people of all ages know, learn, and develop. The recent literature in both fields includes contributions focusing on issues and using theoretical frames that were unthinkable a decade ago. For example, we see an increase in the use of conceptual and methodological tools from anthropology and semiotics to understand how different forms of knowledge are interconnected, how students learn, how textbooks are written, etcetera. Science and mathematics educators also have turned to issues such as identity and emotion as salient to the way in which people of all ages display and develop knowledge and skills. And they use dialectical or phenomenological approaches to answer ever arising questions about learning and development in science and mathematics.

The purpose of this series is to encourage the publication of books that are close to the cutting edge of both fields. The series aims at becoming a leader in providing refreshing and bold new work—rather than out-of-date reproductions of past states of the art—shaping both fields more than reproducing them, thereby closing the traditional gap that exists between journal articles and books in terms of their salience about what is new. The series is intended not only to foster books concerned with knowing, learning, and teaching in school but also with doing and learning mathematics and science across the whole lifespan (e.g., science in kindergarten; mathematics at work); and it is to be a vehicle for publishing books that fall between the two domains—such as when scientists learn about graphs and graphing as part of their work.

How Should I Know?
*Preservice Teachers' Images of **Knowing (by Heart)** in Mathematics and Science*

By

Kathleen T. Nolan
University of Regina, USA

SENSE PUBLISHERS
ROTTERDAM / TAIPEI

A C.I.P. record for this book is available from the Library of Congress.

ISBN 978-90-8790-212-4 (paperback)
ISBN 978-90-8790-213-1 (hardback)

Published by: Sense Publishers,

P.O. Box 21858, 3001 AW
Rotterdam, The Netherlands

Printed on acid-free paper

Dedicated to my mother
and her memory

the most influential teacher in my life

we've noticed that Kathy can't find the missing numerals
we've tried hard to help her
but haven't succeeded as yet
I think it will come to her as she has no trouble
with her additions and subtractions.

(Parent comment, grade one report card)

ACKNOWLEDGEMENTS

There are many people, places, and circumstances to thank for contributions to this book. My gratitude far surpasses what the following few words attempt to convey.

I wish to express my heartfelt thanks to…

Peter Lang Publishing Group for granting permission to reprint the piece entitled, *conSCIENCEness Land(e)scapes* (p. 40). This piece was originally published in Hasebe-Ludt, E. & Hurren, W. (Eds.) (2003). *Curriculum Intertext: Place/Language/Pedagogy* (pp. 71-72).

University of Toronto Press and *Canadian Journal of Science, Mathematics and Technology Education* (www.utpjournals.com/cjsmte) for granting permission to reprint portions of chapter two entitled, *How does light travel?* Material from this chapter was originally published in: Nolan, K. (2003). Casting shadows in the science classroom. *Canadian Journal of Science, Mathematics and Technology Education (CJSMTE), 3*(2), 213-229.

Michel Lokhorst & Peter de Liefde (SensePublishers) and Wolff-Michael Roth (*New Directions in Mathematics and Science Education* Series Editor) for the opportunity to publish this book along with the guidance offered throughout the process.

Family and friends, whose support and encouragement are reflected in numerous ways on the pages of this text. A special thanks to Dad, Art, Aaron, Karen, Kate, Kim, Kiribas, Liz, Lorna, and Paul Hart and my other friends and colleagues at the University of Regina for offering a kaleidoscope of strength, insight and joy on the journey.

CONTENTS

INTRODUCTION [PART I]

A KALEIDOSCOPIC (PRE)VIEW

How to Make a Kaleidoscope

Purpose:
To construct a kaleidoscope

Objectives:

1. To demonstrate properties of light
2. To understand the principle of reflection
3. To explore symmetrical patterns

Background Information:
The word "kaleidoscope" comes from the Greek words *kalos* 'beautiful' and *eidos* 'shape'. Together with the word *scope*, kaleidoscope means 'beautiful shapes to look at'. Based on an ancient Greek mathematical principle, the kaleidoscope was designed in 1817 by Scottish scientist

(Eight people gathered in the lounge for a focus group discussion, March 2000)

Kathy:
As this is our second group discussion and I have already met twice with each of you individually, I thought maybe this would be a good opportunity to explore some of the reasons and/or influences behind your beliefs about math and science knowing.

Just to review some of the things that came out of my first two conversations with each of you... when I asked you to talk about your images of science there were lots of responses initially about science being how the world relates and how things connect to one another. It seemed that these ideas represented what most people thought science *should* be about but then when I asked you to describe what *your* experiences of science were like, they seemed to be different.
Some of the things coming out of our conversations were:

- learning definitions from the text and then later being tested on recall and memorization
- some said science was sort of like a puzzle, with each piece meaning nothing to you by itself but then feeling that you never really got to see the puzzle all pieced together
- the language of the terms in math and science are mostly what you remember but they were usually out of context so you never really

Sir David Brewster. In a kaleidoscope, mirrors reflect light from coloured beads or glass shapes. Beautiful geometric patterns are created which change with each turn of the kaleidoscope. It is a wonder how the view into this magical creation can keep transforming itself into new forms and designs.

Terminology:

Kaleidoscope: a tube for viewing symmetrical, colourful patterns, containing coloured bits of glass or beads and two or more mirrors.
Mirror: a surface that reflects light.
Light: a form of radiant energy that acts on the retina of the eye, by which we see.
See: to perceive, with the eye or the mind.

understood them

- the importance of having answers or knowing where to find them to show that you have knowledge

There are other things too but... does anyone want to talk about one that you particularly relate to?

Roberta:
The definition part for sure. In high school, that's what you did... learning definitions, science fairs, cutting up brains. I'm not sure how they were all put together, but those are the things I remember.

Evelyn:
I know for me, there was never any kind of assessment or evaluation except for tests... and it was just based on your ability to memorize, not your ability to understand.

Kathy:
So you wouldn't be asked to do anything with these definitions...?

Evelyn:
No, just define... the first test of every year... define biology.

Ursula:
I remember that, too. And you know, I hated science and I could memorize the definitions and I could get the marks in the class, but the minute somebody gave me a choice whether or not I'd take science again, I didn't. So obviously I didn't know anything about it or wasn't excited enough

__Experimental Design__:

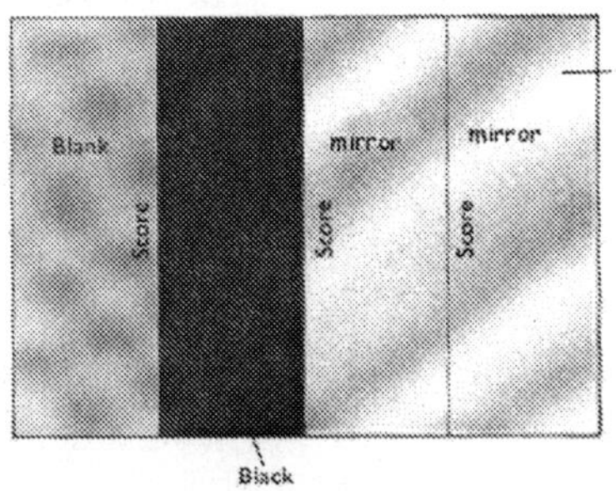

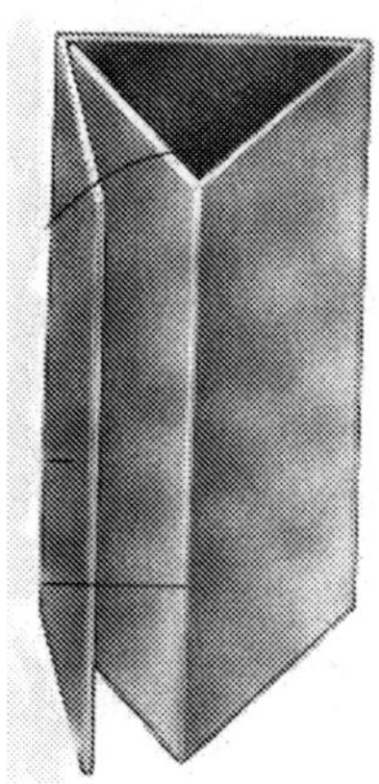

about it to do it on my own. I just memorized what I needed to memorize... for the test.

Roberta:
The only fun part was experiments, at least for me. We were allowed to pick any experiment we wanted to do and present it. That was kind of neat. That was the only fun part.

Ursula:
But did you ever know what your experiment... if it *means* anything? You know, you're doing the experiment but do you know where it fits in [with] what you're doing for the whole year. Like, my science fair experiments were nothing to do with what we did the rest of the year... do you know what I mean? You know, in the biology class, we're looking at things on slides one day and then the next day I'm putting wrinkle cream on apples... to see if it works or not. That was my science fair. Does oil of olay work? (laugh) It stayed all nice and smooth and the rest were wrinkled up. I was so proud of myself. But it had *nothing* to do with what we were learning, not even a little bit.

Roberta:
In elementary school, the examples given in textbooks to do for science experiments— they're generally all materials you can collect around the house. The experiments are very home-based, like they're not expensive experiments. When you go into high school science, you got all the gadgets, you got all the tools, you got... it's almost as if in grade nine you make this

Materials:

2 long, narrow mirrors
sheet of stiff cardboard
black paper
clear plastic
glue, scissors, tape
small coloured glass shapes or beads

Procedure:

1. With the pencil, divide the piece of cardboard into four equal strips or panels. Score the lines of the cardboard so it is easier to fold.
2. Glue the mirrors on two of the panels side-by-side (making sure one is an end panel) and glue the black paper on the panel next to one of the mirrored strips. Leave the fourth panel blank.
3. Fold the cardboard to make a triangular shape, with one panel

transition into 'ok, *this* is the real pristine science world'. You know, 'this is the structure, these are the little boxes, this is biology, this is chemistry...' and it's not as hands-on in the same sense. And I wonder if that has an effect on how students perceive science.

Kathy:
Does it for you?

Roberta:
I was lucky... my teacher did a lot of hands-on stuff and things you could do at home, and I know some didn't have that opportunity.

Harriet:
In grades 3 and 4, I remember we were bouncing balls to see height. That was the science experiment and we did it for two weeks. *Everyday*, bouncing balls... and we progressively moved from smaller to larger [balls]. *To this day*, I don't know what the purpose was. And I actually see the science teacher occasionally and I'm so tempted to go to him, 'I still don't understand'. Stand, drop it. Stand on the chair, drop it. I still don't know.

Ursula:
I remember math, sitting in your desk— even in university, any grade—you were sitting in your desk, and you'd have your textbook open and you'd do the first set, there'd be little instruction part. You'd have stuff that you did together as a class and then the stuff you did on your own in your notebook. He [the teacher] would say

overlapping. Tape this overlapping panel to hold it in place. The angle between the two mirrors should be 60 degrees.

4. Stick a piece of clear plastic over each end of your kaleidoscope.
5. Place the coloured glass shapes or beads over the top of one end and tape or glue some tracing paper over the top of these beads.
6. Your kaleidoscope is now ready. Peer into the kaleidoscope, while looking at a light, and gradually rotate it.

<u>Precautions</u>:

- Be sure the mirrors are at an angle of 60 degrees to one another.
- When taping or gluing the tracing paper on top of the glass or beads, be sure to leave enough room for the pieces to slide about under the paper.
- To see the colours clearly, it is best to look at a bright light as you rotate the kaleidoscope.

'how many of you got that?' and I *never* was there yet. I didn't even have three seconds to think about it, he was already moved on to the next question. It used to frustrate the heck out of me.

Roberta:
I think the way we're being taught to teach kids is more using more hands-on material. So, with our experiences— you've got to wonder how *our* teachers were taught to teach us.

Kathy:
I think I had some pretty good ideas for how to teach when I went out to teach. But when I got busy or when I got stressed or when I got questioned or whatever, I just let the teaching ideas go and the textbook came out and I had to teach the way I was taught. It was easier.

Sylvia:
I don't think I will do that. With the things that I'm comfortable with—the things that we've had lots of time with in science and math—I don't think I'll go back to the way I was taught. But with the things I'm not learning anything in now... then I'll go back to the way I was taught. Somehow, we're still not getting more comfortable with them... it's easier to fall back into that place when you're not having a comfortable experience teaching it.

Kathy:
Do you find yourself asking *why*, instead of just teaching *how* to do something in math? Do you find yourself asking why and what it means?

Calculations and Results:

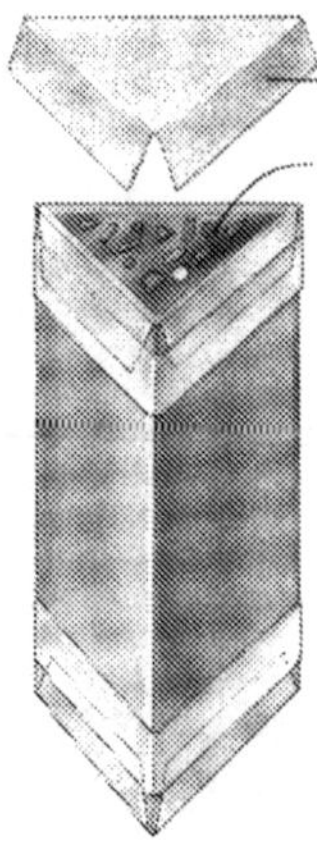

Upon peering into the kaleidoscope, exquisite patterns are seen. Light from the coloured glass shapes or beads reflects off the mirrors to create these patterns. No patterns are ever repeated and all are vivid and colourful.

Conclusions:

1. One of light's properties has been demonstrated: reflection.
2. The principle of reflection is understood, as light reflects off the mirrors placed at 60 degrees to

Helen:
I think I'll be good with manipulatives and making it fun if I'm in K-4, like I want. But if they stick me in like grade 7, I'd go textbook— 'sit down and do some work'—cause like, oh my god, what do you do with grade 7?! I hope that never happens.

Ursula:
I thought grade 4 was going to be scary because... they do things a different way than I do. So, I have to kind of re-train myself. Like, they do things with manipulatives and I just sit there and watch. When I learned to multiply, it was 'carry the one'... tens, *whatever*, you just carry the one, you add it on, it didn't mean anything to me. And they're like lining up these things on their desks. I honestly didn't know how to do it.

Kathy:
The idea of the answers— getting the answer to get the mark to get to the next level; the idea that if you don't know the answer, then you can always find it. Can we look at that a different way? Just because we can get to the answer, does naming an answer mean we have knowledge? What about the other way around: can we have knowledge without answers?

Ursula:
I think you probably can but it would be hard to measure it somehow, especially in older grades. Everything you did had a mark attached to it. So if you had some other way of doing it or some other way of explaining it or thinking about it; it would be a lot harder to have

one another.

3. Symmetrical patterns are explored as the kaleidoscope is rotated; no patterns repeat themselves but all are symmetrically designed.

something to write on your report card. I don't know. And I suppose there's a certain number of facts that you have to know and there's no getting around that. If you want to understand how something works, you need to know that information. But I don't know.

INTRODUCTION [PART II]

TURNING TO THE PARTICIPANTS

"The biggest reason why I thought I would come and talk to you is because I really do have a story with regard to science..."

An elementary preservice teacher in my first interview of my first pilot study spoke these words. They resonated in me as I began to understand the need to listen to the stories of preservice teachers' experiences in mathematics and science. And so the stories begin, as I re-present the stories of my participants mingled with my own stories in this introduction to my research.

"Living, telling, retelling, and reliving mark the qualities of a life." (Clandinin & Connelly, 2000, p. 187)

One of my participants, Roberta (named after neurologist and astronaut Roberta Bondar), first introduced herself to me in an e-mail message, sent shortly after I gave a brief presentation in her middle years science class. She indicated that she would be willing to participate in my study, saying: "I have a degree in music and faced many of the same questions about gender during my [degree]." It was apparent that, while my presentation spoke to many other issues of interest in my research, the one she heard and identified with most clearly was gender. Our conversations remained mostly on this gender path, with Roberta informing me that she consistently fought gender stereotypes all through school in many different subjects.

Roberta also spoke often about her opposition to the perpetuation of the high status afforded mathematics and science. I remember very clearly the determination, and often anger, in her voice when she related stories to me about being forced to take math instead of music in her senior high school years, even though it was music that she knew she would study at university. Roberta found it very difficult to understand why there were not more efforts made to develop and teach interdisciplinary, holistic curricula, rather than creating discrete learners who "go out into society with a full brain of facts" (Roberta, October 1999).

I can still remember my (first) Ph.D. proposal, written to accompany my application to the university's graduate program. Looking back on it, it is clear that the proposal does not reflect this research study. However, it was based on my own lived experience and personal desire to address the problems I per-

ceived with mathematics and science education, especially for girls, and especially once students enter middle school classrooms. In my own schooling, I had experienced many boundary lines between subject disciplines— lines which seemed natural to me at the time since both textbooks and class schedules certainly portrayed subjects as discrete. I somehow knew, at the time, that real life crosses and smudges the boundaries between subjects, even though I had not experienced such boundary crossings in school.

Later on, as a teacher, I tutored girls in middle school mathematics and science. They would tell me that they did not understand the subjects and they never seemed to be able to see any connections to real life. "I'm just not very good at math", the girls would tell me before our tutoring session even began. I was certain that these comments were deeply rooted in the way the subjects were portrayed. Like parallel lines running along side one another all day long, separated only by the physical distance between classrooms, school structures made it apparent that subjects were not intended to cross.

So, I decided that my doctoral research would be a problem-based exploration into mathematics and science integration for preservice teachers. It seemed like a great idea at the time— a perfect solution to a problem that I thought I knew very well. And then I visited a physical science workshop for elementary preservice teachers on the atom. When the instructor asked the students to discuss their ideas about why there is a need for an atomic model, a barrage of comments and qualifiers echoed throughout the laboratory. [A collage representing a few of these comments appear on p. 10.]

From this moment on, after hearing such expressions of non-knowing, the focus of my research turned to questions relating to perceptions of knowing and how preservice teachers understand what it means to know (in) mathematics and science.

One elementary preservice teacher from this science education class, who volunteered for my study, was Ursula (named after physicist Ursula Franklin). She had written a paper, only days before I described my research to her class, in which she summed up her attitudes toward science at the start of her science education course:

> I hate Science, truth be known, or have always believed that I do. What is there to like? It is intimidating—full of facts and figures, charts and graphs, and theories that attempt to bend my mind into places it is reluctant to go! At the onset of this course, I was already sure that I was destined to fail. I could not "get" Science in my earlier education, and had similar problems in university. And if I can't get it, how on earth will I teach it?! (Ursula, October, 1999)

Participant observations in a science education workshop[1]

Why is there an atom?

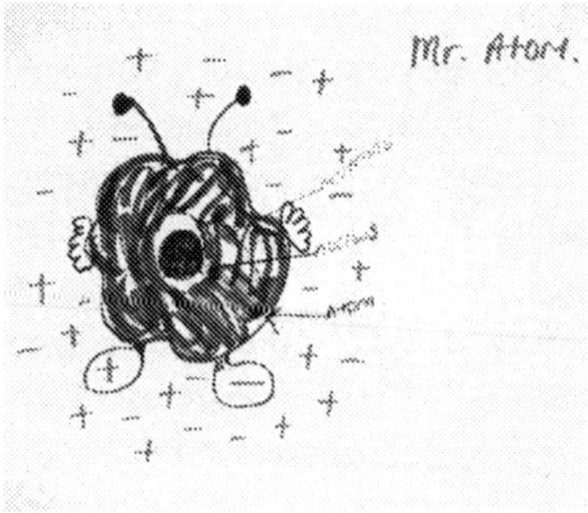

I don't know anything about this.

Isn't it because…? Oh, I don't know.

I've taken Physics at university and I don't know this.

I'm really clueless on this. Sorry.

I don't remember anything. I failed science.

Instructor: So what did you decide…?
Preservice teacher: We decided we didn't know anything.

> Science, throughout my education, has meant memorizing someone else's ideas. I never had any input into what I was learning, and was never provided any answer as to where this "life-changing" information fit into my real life! I feel that I have been condescended to, and preached at, but have never really had an experience in which I "learned" Science. (Ursula, October, 1999)

From the perspective of the importance of lived experience in my research, I find the excerpt above to be a very interesting illustration of the natural movement of narratives within the temporal dimension of experience (Clandinin & Connelly, 2000). The importance of experience, as infiltrating and weaving throughout past, present, and future, is illustrated in Ursula's story above. Her past school experiences contribute to her present anxiety about the science course she is about to begin, and the panic she feels about her future, teaching something she just did not "get", or learn, herself. The story also illustrates the dynamics of an interactive dimension of narratives (Clandinin & Connelly, 2000). She expresses the contradictions between her inner feelings of a destiny to fail and of having little value attached to her own ideas and her perceptions of the outer environmental constraints of curriculum and teachers that provided no choices and asked for no input.

In writing about a three-dimensional space of narrative inquiry, Clandinin and Connelly (2000) state that "as inquirers we meet ourselves in the past, the present, and the future… we tell remembered stories of ourselves from earlier times as well as more current stories. All of these stories offer possible plotlines for our futures" (p. 60).

the future holds
all dreams
expectations and *possibilities*

try not repeat
the present the past allows
time to *experience* reflection
claim our now to *ponder* images
and know in the rearview mirror
how quickly but it must not
we move into interfere with
reasons to avoid

the future holds
all
dreams

I believe that my participants, as inquirers themselves in my research study, benefited by reflecting on the role of storying their experiences (and experiencing their stories) in their future as teachers.

Ursula shared many stories with me and she was very passionate in her concern for how these past school experiences might influence her teaching. She was mindful of the fact that it would be very easy for her to fall into the 'trap' of teaching as she was taught, despite how vehemently she opposed giving students similarly "meaningless experiences". At one point Ursula told me that she was scared to teach mathematics and science because she experienced the learning of these subjects one way—from a textbook. Her only memories of science were of reading the textbook, memorizing definitions, and then "spitting them back out on a test".

It scares me that I have to learn it sufficiently to teach it to someone else.
It's a little different than I need to learn it to pass a test.
I guess it *is* a test, but it's a more important test.
(Ursula, November 1999)

Like Ursula, I have also had my share of school experiences that taught me what it means to know and to learn science…

I called my sister the other day, just to reminisce and share school stories. She is two years older than I am but we attended the same schools, had the same teachers, used the same textbooks, and, as it turns out, even had a few identical notebooks. Actually, I should say 'have' identical notebooks, as we have saved some of them to this very day. We both agree that our grade nine science notebooks are true works of art (in need of framing or embalming or something)— that is, if you consider direct copies of textbooks 'works of art'. Still, even though no ideas, words, or pictures in those notebooks are our own creations, we are very proud of them just the same. How often do simple notebooks encapsulate everything (and I mean *everything*) that was done in an entire year of school science?

To this very day, we can still see our grade nine teacher sitting at his desk, and hear him telling us that we were about to embark on an entirely new chapter of our lives (in our textbook). He would carefully outline, page by page, which diagrams we were to draw and colour, which "experimental write-ups" we were to copy and study, and how many days or weeks we would have to complete this sciencing exploration into the depths of the textbook pages and our Crayola boxes. I think we both learned many important things from those grade nine days of sciencing. I know I certainly did.

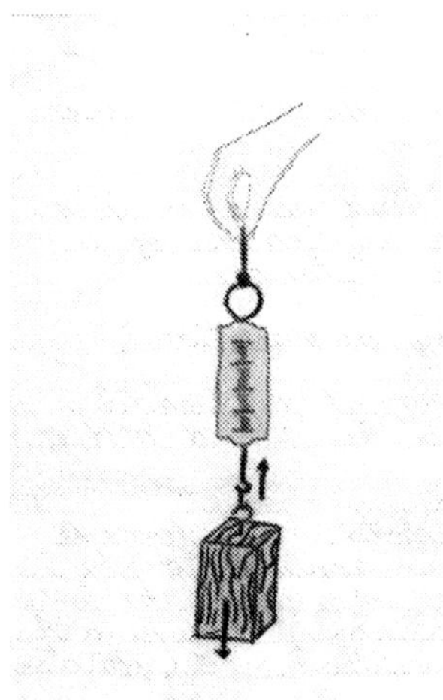

I learned the significance of **science experiments**: We were to copy those experimental write-ups word for word from the text, being very careful not to miss any steps lest our notebooks would represent failed experiments.

I learned the significance of **objectivity in science**: Not only were all the students in my class completely protected from the laboratory (was there one?) and possible experimental errors, but my drawings illustrated how minimal subjective 'interference' could be realized by having only hands involved in any experiment… accurately representing a truly disembodied science knowing.

I learned the significance of **teacher demonstrations in science**: My sister would proudly report at home how *her* accurate drawings and precise colourings were held up by the teacher in front of the science class for all to observe, and maybe even for all to hypothesize how they might get their notebooks to look as lovely.

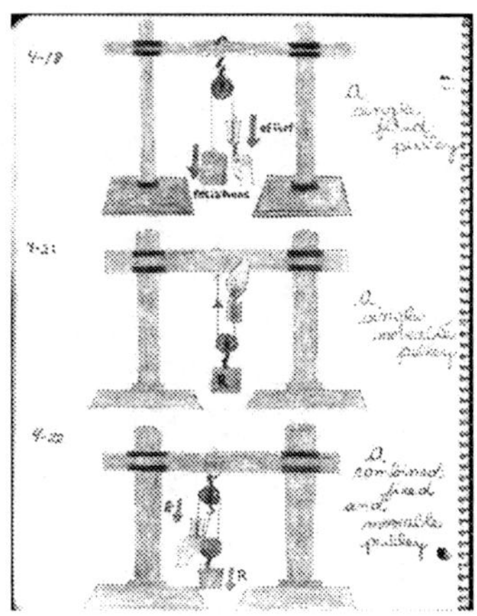

I learned about the significance (and abundance) of **abstract concepts in science**: I was certain that my grade nine drawings labelled 'levers' and 'pulleys' only represented *ideas* of things… otherwise, we would have seen the real thing in the classroom or identified them as real-life concrete objects that one could touch and play with, right?

I learned the significance of **personal knowledge construction in science**: I knew it was not a sufficient learning experience to merely *view* textbook diagrams and *read* textbook words; we needed to personally experience the construction of these diagrams and words on the pages of our very own notebooks.

I learned the significance of **symbolic representation in science**: Using the letters V, R, and I to stand for the longer words of voltage, resistance and current made my sciencing go a lot faster and smoother when copying solved problems from my textbook to my notebook (even though I always assumed that **I** for current was a funny typo that the textbook authors never corrected).

I learned the significance of **experiential learning in science**: *Crayola* makes several shades of grey and unless I tried all of them out for myself, I would not have learned which one was perfectly suited for the little triangle under the brown stick in the thing labelled 'lever'.

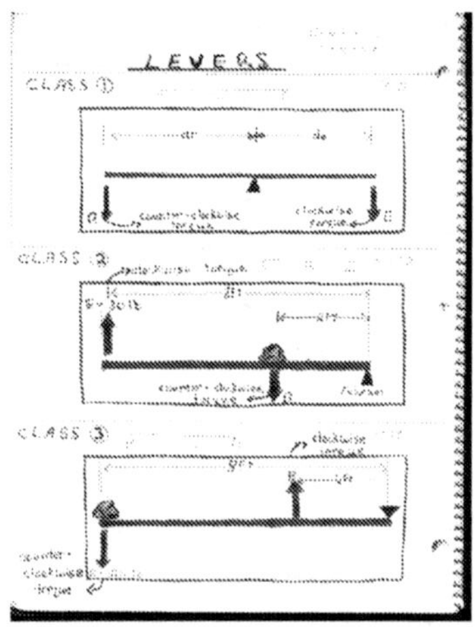

Yes, my sister and I often share a laugh or two over those days of grade nine science. That is,

... until we speak about our grade eleven physics classes, when the teacher used the word 'lever' over and over again and we could only nod and smile in agreement that, yes, the word was very familiar to us; we never really knew how to communicate to that teacher that the word held no meaning for us whatsoever outside of the 'natural' context of where we learned it— from our lovely grade nine textbook picture labelled 'lever'.

... until we speak about my embarrassment, many years later as a high school physics teacher, being handed an inventory *list* (that is, labels without pictures) of physics equipment and asked to go through the lab cupboards to count and check-off items—an activity that my principal indicated would take about two hours, not six hours of frustration, tears, and feelings of being an imposter.

... until we speak about my sister's 13-year old daughter, currently puzzled by how she could possibly value her own words and lived experiences in grade eight science, as she is caught in the textbook current of looking at pictures, answering questions, writing notes, memorizing terms, and performing tests.

I was not the only participant in my research study who had such dominant textbook and colouring introductions to science. Maude (named after chemist Maude Menten) also shared some personal notebook collections of her early science experiences with me. She laughed as she pointed out to me the teacher's red ink strewn across her drawing of "The earth and its neighbours", indicating that it should be coloured. Maude was an after-degree student, having previously completed a degree in nutritional science and working a few years in the field. In conversations with Maude and other participants I was generally curious about how and why school mathematics and science became so answer-focused. Mathematicians and Scientists are forever asking and exploring "what if..." questions without arriving at absolute answers, but instead constructing tentative models and theories. Yet, in my conversations with these preservice teachers, it was always difficult to perceive the idea of someone having knowledge without having answers. "Yeah, knowledge and answers go hand-in-hand", Maude commented.

Maude spoke to me about gradually becoming more comfortable with not having the answers herself, but it was still clear that, for her, it was important to know that the answers existed *somewhere* and that she could always find them for herself and her students. Could she imagine teaching and learning mathematics and science without answers? Could she imagine asking ambiguous and unanswerable questions in math and science?

Recently, I was employed as a substitute teacher in a mathematics and science adult education class. The students in the class were all aiming to complete their grade twelve equivalency; they worked independently, at different subjects and levels within the classroom. Most students in the classroom, however, had one common goal: to get through the necessary steps to obtain the credits, to secure the job or promotion.

The students.
Nine in number.

One graphs systems of equations on a graphics calculator and wonders why he wasted all that time and graph paper in school.

Another ponders why $(-6)^2$ is not 12.

A man draws a blank as he flips through the pages of the biology chapter, seeking the right words to place in the questions' blank spaces.

Hands on chin, she says it's been almost a year since she's done this math, but it'll come.

From under the baseball cap, his head never leaves kinematics and I wonder why he has no questions for me.

The other kinematics student needs to know if she's supposed to just READ about ticker tapes, since she certainly can't be expected to DO the experiment, right here, right now.

The man with the cap and plaid shirt, breathing heavily as he returns from writing a math test, is clinging to a possible 50 as he says 'some of the stuff... just gone.'

The most advanced student has circular function angst over the inverse of inverse sine functions, as if he'll ever use them.

And the man at the back table calls me over every so often to interpret what they [the nebulous textbook] want for an answer, as his just aren't matching those in the back of the book.

And it's a room full of answer-want
It boosts my self-confidence to give them what they want
It pumps me up to be able to take away some of their frustration
And then I return to the pages of my own research
where I am trying to understand
why answers are important
and what it all means
to have knowledge
why it matters
if I'm tested
on what
I know
(not)

In my research, it is always difficult to articulate my anticipated results or 'answers'. Early in my study, I could just say something like "still in process" or "caught in the thick of things" and "can't see the forest for the trees", and other applicable clichés. I gave a lot of answers that essentially attempted to dodge giving answers. As my research progressed however it became more difficult to express how I was trying to *not* be answer-oriented, since (I explained) this is a positivistic goal that is ingrained in so many of us and would we not be more enlightened if we thought of better questions to ask instead of closing things down with attempts at absolute answers? (Dodging again, eh?)

Feeling shadowed by a perceived need for answers, I have learned that it is not so easy to explain or give labels to what my research is about. Depending on with whom I am speaking, I describe my research in very different ways. To say my research is about gender differences or gender issues in mathematics and science is a simple, yet sometimes dangerous, response. Such a neat and tidy description might give the impression that there is a neat and tidy solution that goes with it. During the course of my research, a high school teacher received word that I was a Ph.D. student at the university researching 'gender issues in math and science'. He called my office and left a message that he'd like to speak with me. I thought, "Great! A teacher in the field is interested in my research" and so we proceeded to play phone tag, never quite catching up with each other. Finally, after two or three messages each, he left a message saying that since it was difficult to reach one another, he would just leave his school address and maybe I could send a package of materials for him to use in his classroom.

My first reaction to this request was to think: "Oh my gosh, I have no package! What am I doing with my time? My research is useless".

I guess it is natural to want practical applications that will alleviate inequities (or other perceived 'problems'). I tend to think, however, that the real issues to address in education are the deeper issues that involve a lot more work, first in even noticing them and then in continuing to notice and address them. There is no panacea (or package) for the deeply personal and political issues.

I dodged having to explain myself and my package-less research project by not returning the teacher's call or mailing him anything. How can I tell high school teachers, who work day-in and day-out in the practice of curriculum, that I think it is important for preservice teachers to reflect on questions and issues related to the very curriculum that teachers in the field seldom have time to even *think* about, as they attempt to *cover* it?

Some crucial questions and issues that necessitate reflection are represented in what I refer to as my research constellation inquiry space—a space of exploration that is flexibly (re)defined and (re)shaped by influences of gender, culture, and context. "A constellation is a juxtaposed rather than integrated cluster of changing elements that resist reduction to a common denominator, essential core, or generative first principle" (Jay, as quoted in Bernstein, 1991, p. 8). Use of the constellation metaphor highlights the complexity and multi-dimensionality of my research project without trying to force a common denominator or essential core. The metaphor also suggests that the clustering or juxtaposing of these research 'elements' illustrates how the questions naturally influence, and shed light on, each other throughout the research process. The constellation reflects the dynamic nature of these influences, as they continually shape and perform in the space through increased understanding about…

My research constellation of inquiry[2]…

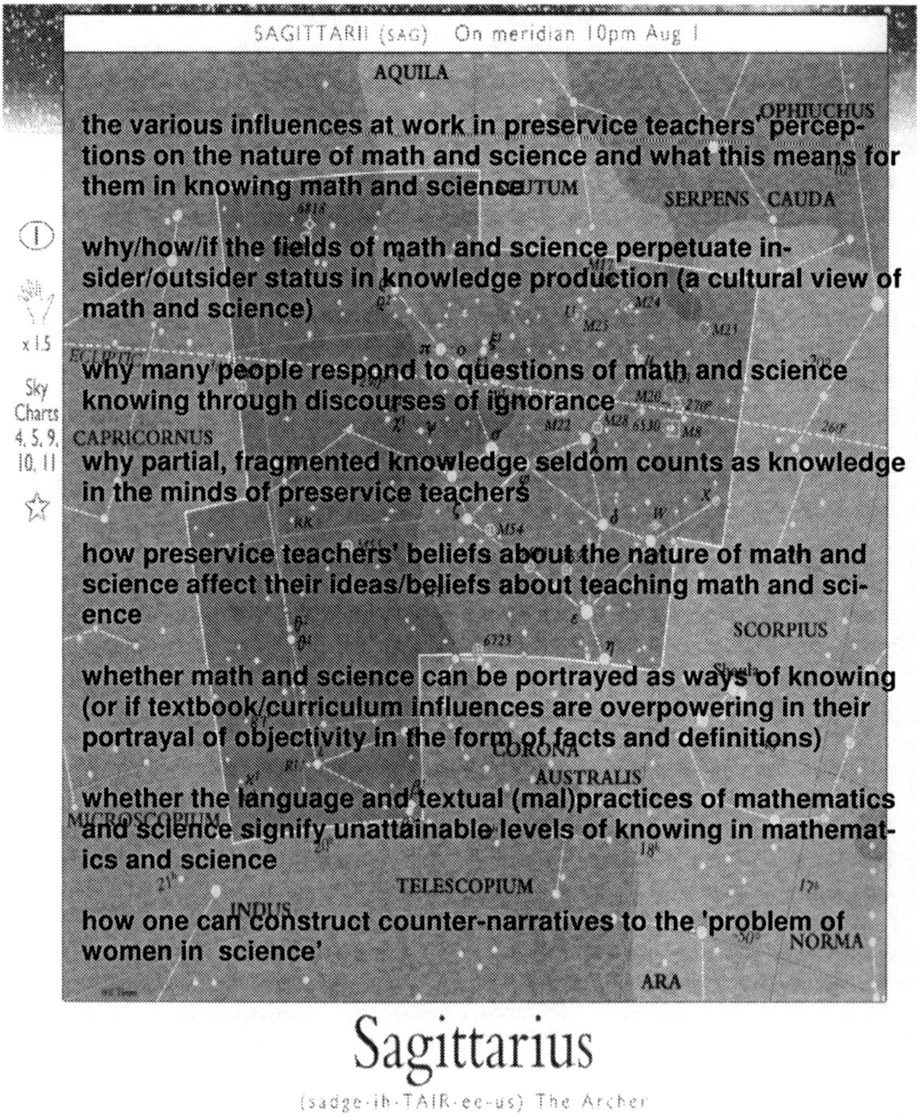

The constellation space glitters with many 'why' and 'how' questions— questions which I know do not have absolute answers. Yet, they are important questions to critically ponder and explore through participant experiences.

I asked one of my participants, Elsie (named after aeronautical engineer Elizabeth "Elsie" MacGill), if she spent a lot of time wondering 'why' in science. Like myself, Elsie always succeeded in doing well in mathematics and science, in performing the necessary tasks and understanding very well how to *perform* these subjects. I wondered though if she felt challenged by wanting to know *why* something is done, rather than just *how* something is done.

With some things, I was interested enough
to care why it worked.
I remember doing levers—1st, 2nd, and 3rd class levers—and she
[the teacher]
would have the diagrams up on the overhead
and I remember wishing we could play around with this.
(Elsie, November 1999)

(It is ironic that a common NON hands-on experience in science is with levers, especially since much of our life-worlds are hinged on these objects.)

Elsie presented an especially interesting perspective about women in math and science. In our first conversation, Elsie talked to me about changes in her career directions over the past few years. She had always thoroughly enjoyed math and science in high school and decided to enroll in an engineering program at university. We did not speak very much about why she changed her mind part way through the engineering degree (transferring to the Faculty of Education) except for her "lack of comfort with the people". We did speak at great length about her high school experiences and why she enjoyed math and science. It appeared that Elsie and I had very similar experiences and we even confided in one another about our reasons for the educational choices we made as a result of these experiences. She was always very 'good at' mathematics and science and enjoyed the challenges presented by these subjects. We confided in one another about how we enjoyed the challenge to prove that we could 'do' mathematics and science. We were challenged by the typical educational influences of demanding teachers and high-grade goals, but we also admitted to secretly enjoying the challenges (and special attention) that came with being in a mostly male environment. I think this highlights some very interesting contradictions in issues of gender equity and feminist theory.

It is a strong belief of mine (and a concern in my research) that to focus on girls and women in mathematics and science *as an anomaly* is to perpetuate the notion that it is natural to be surprised by women choosing to study math and science. This, in turn, perpetuates different expectations for women. Traditionally, mathematics and science gender equity initiatives have focused on making girls and women a better fit for math and science. The problem with initiatives of this nature, however, is that they risk reifying women as *objects of a problem* that needs fixing. In other words, I think too much time and energy is spent engendering gender in some of these initiatives.

Ironically, and in apparent contradiction to what I just wrote, I always *liked* the attention of being different— of being a woman in math and science. There is a satisfying feeling in generating surprise in people, and then in realizing that the surprised people automatically think I must be smart, without even talking to me about math and science!

Helen (named after astronomer Helen Hogg-Priestley) began her participation in my study after having reflected to some extent on her own personal experiences in math and science classrooms, as part of a course in women and education. She writes:

> We were having a discussion in one of the groups… and many of us were saying that we did not like math class in high school because we would always want to know ***why*** the teachers were doing something, and they would just say that "that's just the way it is". We discussed how frustrating this was, and still is. I have always had this problem, in every single math class I have ever taken. When I am having difficulty, I always approach the teacher for help, and every single time, the question I have is: "Well, WHY do I have to do this here?" or "I don't understand WHY you did this step on the board." Some of my more understanding teachers took the time to do their best to explain it to me, but some simply became frustrated. Every one of my math teachers since Grade 9 have been male, so that might have something to do with why many of them did not sympathize with my need to know "why" (Helen, April 1999).

Initially, Helen did not volunteer to participate in my research study. She offered me a paper she had recently written, wondering if it might be useful for me to read and use in my research. When I asked her if she would be interested in being a participant, she was a bit reluctant at first, saying that she wasn't sure how/if she would be of any help to me. I am pleased that Helen decided to join the research team, since I think we both learned a great deal from our conversations together. As I journeyed along with each of my participants in this study, I experienced many moments in which my own personal experiences as a student mingled and meandered in and between the stories of each of these eight women (much like I am re-presenting the stories in this introduction). We shared feelings of confusion, frustration, anger, and insecurity as we reflected on our student experiences of mathematics and science knowing. But with Helen, perhaps more so than with anyone else, our conversations drew me into reflections on my own personal experiences as a *teacher*, as well as a student.

'Helens' appeared in every one of my mathematics classes in every one of my teaching years, and I seldom acknowledged it. They were often the students who sat and worked quietly, occasionally wanting to know 'why' but seldom demanding anything more of my teacher time and energy. As I reflect on these experiences, I am quite certain that I must have convinced myself at the time that when the 'whys' stopped being asked by my students, it meant that they were successful learners and I was a successful teacher and I could go home patting myself on the back. I preferred to believe that the 'Helens' stopped asking 'why' because they understood and they learned, not because they became self-defeated. I could not have lived comfortably with their self-defeat; it was much easier to carry on if I imagined that I was doing my job better than that.

It is not just because I am interested in this research that I find myself reflecting on these teaching issues. It is because I found myself reflecting on these teaching issues that I became interested in this research. I have been aware for some time now that, as a teacher, I spent a lot of time surviving one step ahead of the students in 'learning' the material, but two steps ahead in the denial of it.

To understand more about 'why' I am interested in my particular research questions, it is important to set the context related to my own lived experience. To this very day—after completing every science and math course offered at my high school, a Bachelor of Science degree in math and physics, a Master of Science in physics, eight years of experience teaching math and physics—after all of this, the hairs on the back of my neck stand on end and I break out in a cold sweat when I think someone is about to ask me a question that tests my 'knowledge' in math or science. It's as if I'm expected to have some understanding of these disciplines after studying them and being immersed in this culture for approximately twenty years! Anyone who has learned the rules and played a game very well knows that to play the game doesn't necessarily mean one has a practical understanding of the meaning or development of the rules, or in fact that one has any ability to apply these rules in another context (should they need or want to). Living within the confines of rules about what works and what doesn't work in a given context does not imply any expertise other than that which is demonstrated at face value. I learned the rules of operating within the mathematical science culture, and within that context I could apply the rules quite well. Why is it that so many people assume so much more of my 'knowledge'?

The power of deciding what counts as knowledge and what knowledge we cannot do without, has served to marginalize many groups of people. Having phrases like "if you don't know this …" hang over one's head and heart as a threat, leads to people trying to cover up exactly what it is they don't know or, at the very least, to claiming ignorance and inexperience right away before being subjected to a 'knowledge test'. (Personal journal entry, October 1999)

I now think that if I had taken steps toward who I really wanted to be as a teacher, instead of always trying to "maintain a strong front" (in the words of my participant, Evelyn) in sidestepping an admission to not knowing 'why' or 'how' or 'if', then there may have been more authentic teaching, learning, and questioning occurring in my classrooms.

Evelyn (named after mathematics ecologist Evelyn Pielou), a prospective middle years mathematics and science (among other subjects) teacher, taught me a lot about the power of the *pressure to know* as a teacher. It was very apparent from our first meeting that Evelyn was quite capable and willing to reflect on her mathematics and science experiences at a deep level. We regularly reflected on the contradictory messages she perceived in the knowing of math and science. Before I began my research project, I had studied just one introductory course in qualitative research methodologies. Being rather skeptical of research processes, I envisioned my role as a researcher to be about gathering interview 'data' and then cloistering myself off somewhere to produce an interpretive and/or critical and/or feminist narrative piece of work. Therefore, I vowed early on in the design of my study that I would make both the 'data collection' and the 'interpretation' processes as collaborative and participatory as possible. My conversations with Evelyn journeyed quite a distance along this path, and in a naturally emergent manner I think. As an example, consider the following conversation excerpt, where Evelyn turned the tables somewhat and began asking *me* the questions.

Evelyn: Do you find it frustrating? You come from a science background *and* you're an educator. Is it frustrating for you, as an instructor, to see people coming into your classroom not being excited about science?

Kathy: A little bit, *but* I can understand where it comes from because... when I first started teaching, if I wasn't comfortable with my knowledge in something then I would skip it or I would learn the parts that I wanted to... I would limit the students' explorations in the area by just giving them what *I* was comfortable with. And sometimes, yeah, it would be given kind of dryly. I couldn't always start with their experience because I didn't **know** how to teach that way; no-one had ever done that with **me**. My experience was: let's learn the formal science and then, at the end of the chapter when we do questions, we'll see some examples. Like learning about parabolas in math— I learned everything about their equations and graphs, and **then** maybe I saw an example of where they exist, like in a volleyball serve. Only then did I start to get a picture of what they were. I don't remember ever being taught from the perspective of the problem or the experience, so it was hard for me to think that way in my own teaching.

I enjoyed my conversations with Evelyn, partly because she shared some critically reflective thoughts with me on how she never felt welcome or invited to participate in the culture of science—the science 'club' as she called it. She demonstrated to me, in many ways, that it is a big step to make between conscientization (becoming aware of the need/desire for change) and actual change. Through her storied reflections, contradictions, and metaphorical language of rebuses and puzzles, Evelyn expressed an astute mindfulness of the hegemony of school science. Together, however, we were often at a loss to move beyond mere commiseration on the state of affairs to a reflection on the transformative possibilities.

If I learned anything during my research process, it is that everyone has a story (or two) about their school mathematics and science experiences. When asked what my research was about, I usually fumbled my way through various descriptors—math, science, knowing, experiences, gender, answers, 'good at', etc. etc. Once I had put together a coherent sentence from within my state of research immersion, I always eagerly anticipated the person's response. And there *was* always a response. Once in a while perhaps, someone would respond curtly with 'oh, that's interesting', which I took to mean that s(he) would really rather not talk about it any further, and there was no story forthcoming. Most of the time, however, the person would voluntarily transform my brief research description into a testimonial of what knowing (in) math and science meant to her/him. But that makes sense if one thinks about it for a moment. In Western societies, the experience of 'studying' mathematics and science is one shared by the majority of the population. Of course, there are lots of people who claim to have no memories whatsoever of school mathematics and science, and then there are those who dropped it the first chance they got, but all in all, I think the topic of mathematics and science represents common ground for conversation (or commiseration, as is often the case). I discovered multiple milieus for conversing on my research topic (and the topic of school math and science in general). It makes for a great cocktail party heart-to-heart, or for passing the hours in the *Safeway* cashier queue, or as a distraction during troubling aircraft turbulence at 30 000 feet. Or even, as it so happened, when bargaining for the best cookie deal at the local bakery…

My request to mix cookies, obtaining equal numbers of each kind,
was fairly straightforward, especially with the combination of 5 kinds
and 40 cookies in total.

The task of calculating the cost was not as enthusiastically embraced,
however. The dozen and half-dozen quoted prices lent themselves
more readily to customers abiding by multiples of 6.

It is never my intention to draw attention to a person's mathematical
(dis)interests. So, when I volunteered a method for figuring out the
cost, I felt the need to pardon my interference.

"Don't mind me, I am a math teacher, and I'm always problem solving."

The cashier looked surprised. "A math teacher?" Before I could nod
and shrug my shoulders in casual agreement, she added,

"I've never been very good at math."

"To this day", she reminisced, "I can still remember sitting
at the kitchen table, doing one math problem after another.
I guess that's the only way to learn it— to do them
over and over again.

But I was never very good at it."

My participant, Harriet (named after nuclear physicist Harriet Brooks Pitcher) talked with me about *her* feelings of not being good at math and science. She reflected in particular on the gendered messages she received loud and clear in her math and science classrooms— messages that outlined for her who was 'good at' science. She explicitly remembered how her enthusiasm to participate in science diminished ever so dramatically when she noticed that the boys were always picked for the demonstrations, despite the show of hands from the girls volunteering to participate also. In our conversations she expressed dismay and disbelief that, after being so critical of those gendered messages in her own school experiences, she found herself actually picking the boys more than the girls to answer questions in her pre-internship classroom. She was struck by this contradiction and quite dissatisfied that she had fallen into the 'trap' of teaching as she was taught, in the very roles of which she was so critical. Unfortunately, I do not believe she is the only unwilling and unaware participant in the perpetuation of these subtle gendered messages in the classroom.

Just last week, I enjoyed the opportunity to work with elementary school teachers during a professional development science workshop. Many hands-on activities were available for the teachers to experience and ponder. One particular station required the teachers to mix together cornstarch and water in a bowl and then to use their hands to get right into the bowl to play with this mysterious solid/liquid substance. Approaching a pair of teachers who were very enthusiastic about the potential for such an activity in their classrooms, I overheard one of them exclaim: "Oh, the boys in my class will love this!" When my retort of "what about the girls?" was ignored, I thought about asking the same question of her tall students, her blue-eyed students, or maybe even her favourite students. (Personal journal entry, October 2000)

In our conversations, Harriet reflected often on some vivid memories of particular significant influences in her life. She remembered learning that her parents never liked science and never felt they were very good at it, but that they never spoke even once about it while Harriet was in school because they did not want to influence *her* like or dislike of science. Harriet also spoke about the influence of her teachers…

> Even in grade 4
> you still knew what your teacher liked and didn't like
> and she did not enjoy doing science.
> (Harriet, November 1999)

and about the science textbook in her early elementary school years…

> I don't know if you've seen them
> They were pink science textbooks
> Oh, they were so archaic
> They were big, stinky, pink ones
> They smelled like that old library kind of smell
> It was awful
> And we'd read these things...

Speaking of influences. I am fluent in being influenced. When my high school teachers and guidance counselor directed me away from my adolescent dreams of being a photographer and toward a university degree in the sciences (because I was so good at these subjects) I concurred that my ninety-something average in math and physics would surely "go to waste" if I did not use it. It is as if I thought I would be taking photographs of the ink fading on my report cards. (Personal journal entry, October 2000)

My participant, Sylvia (named after medical physicist Sylvia Fedoruk) shared a great deal about her discomfort with her knowledge of/in science. She expressed *some* comfort in being able to demonstrate her knowledge limitations/boundaries to her students ("as long as they are K-5 students"), but she was very fearful that parents and/or older students would test (and be disappointed by) her knowledge of science. She spoke about "lacking something, but not sure what."

> What if they ask something of me that I can't answer?
> What if I look stupid in front of them?
> I could never teach high school.
> (Sylvia, November 1999)

This is why, she explained, it was easier for her to dismiss her science knowledge "up front", as a safety net against "being found out" to be wrong. She spoke of qualifying both her questions and answers in science with "I don't know but…"

When asked about her images of mathematics and science (and knowledge in/of these subjects), Sylvia used the expressions “out there” and “like a winding path, with forked roads, and you're not really sure where to go”. She also shared her struggle against images of bald men, lab coats, and thick glasses as the people who *do* science. She knew that was not really what science is about, but, for her, the image would not easily fade.

Perhaps it is easy to fall into ‘traps’ of teaching as we were taught, even if we are critical of those very same teaching and learning experiences. Theories of constructivism teach us that learning new ideas and having new experiences do not simply displace and replace old ideas and previous experiences. This is true for images of mathematics and science *and* for the familiar structures defining what counts as knowledge in mathematics and science. For me, and for my participants, math and science were presented as challenging subjects, not for the feint of heart (or mind). I have experienced ways of knowing (in) mathematics and science as limiting and often exclusionary, yet I have often perpetuated the very same beliefs and values in my own teaching.

"I'm flustered!" she exclaimed as she reluctantly handed in her test on mathematical functions.

"Yes," I said reassuring myself, "a test is certainly not the time you want to be encountering a problem-type for the first time. My tests are designed to see what you have been doing all along and whether you have practiced enough." Disgruntled, the student left the classroom to commiserate with her classmates just outside the door, likely measuring their heart rates before sprinting off to their next exercise in futility.

For the first time I can remember, doubts began to creep into my mathematical mind (and heart) as I gathered the tests together and left the classroom. Normally, at about this time, I would pat myself on the back for a test well made— a test that reflects the undeniable challenging nature of mathematics; a test that separates the 'haves' from the 'have-nots', if you know what I mean. I am being facetious of course, since *I* do not even know what that means, except for believing that it symbolizes unmistakable mathematical arrogance. Just because *I* survived my school days on the privileged side of this binary opposition, does not mean it is the only way to 'know' math. I started wondering how I could continue to comfortably contradict myself, as I *preached* math for all and *practiced* math for the few.

And now, I suppose, I have come full circle in this narrative cycle of introducing my participants and myself, as a way of introducing the lenses through which I (re)present the following text. I am constantly mindful of the fact that as I choose to write certain stories and interpretations, I am (un)consciously choosing not to write others.

Theory resides in the spaces in/between these stories, just as countless stories reside in the spaces of educational research theory, though they are not often explicitly told. I have chosen to present my research introduction in storied form, to provide opportunities for the reader to experience "narrative resonance" (Conle, 1996) with some of the stories and to choose to let other stories go. The stories are also re-presented to encourage the reader to reflect on her/his own experiences before reading more of this research text that explores the experiences of these eight research participants.

We learn from experiences. But, like dreams, the experiences do not have meanings embedded within—we constitute meaning in how we remember them, talk about them, interpret them, embody them. It is then, in a sense, the *discourse of experience* (Britzman, 1990) that is evocative.

> … experience does not "tell" us who we are or what we see: we are the tellers of experience. And yet our potential to retell, set by the conditions of discourse, is always transposed by our own history, deep commitments, and normative notions of what constitutes truth, power, authority, and knowledge. (Britzman, 1990, p. 26)

By reflecting on and sharing one's own discourse of experience, meaning, understanding, and learning all take shape and direction. Is it not worthwhile then to explore preservice teachers' discourses of experience?

WHAT IS SCIENCE
CH. 1
postMODERN conSCIENCEness: Reflections on Light
Refraction
IDENTITIES
VERSES
PROPERTY
EQUALITY
(definitions)
virtual imag
focal leng
real im
effort
resistance
R

postMODERN conSCIENCEness

Reflections on Light

Overview In this preface, I would like to discuss some of the features of this (inner) research 'text' that I have entitled *post-MODERN conSCIENCEness: Reflections on Light.* The title and format of this inner text have been chosen as a parody on physical science textbooks. Through conversations with preservice teachers, I learned how a school textbook (its over-dependence in classrooms and its style of portraying science knowledge as unchanging) can be a major influence in the 'learning' of science. The word conSCIENCEness has been created to convey the idea of "being with science", but hopefully in postMODERN ways. One possibility for imagining postMODERN ways might be to think of this text as an exploration into the many narratives of knowing (in) math and science. There is not one grand narrative that can be written to enable a reader to understand how the experiences of these preservice teacher research participants have shaped their images of knowing, including what counts as knowledge and whose knowledge is of most worth. The text introduces and explores several different stories of exclusion from the cultures of school mathematics and science. In this sense, then, the only grand narrative that can be imag(in)ed is one in which the personal lived experience narratives of my participants mingle and interweave to create a sort of kaleidoscope of narratives. In fact, the narratives might best be viewed as a kaleidoscope of "counternarratives" (Giroux, Lankshear, McLaren, & Peters, 1996, p. 2) since the voiced experiences of the preservice teachers run counter to and challenge the official, grand narratives.

With each turn of a kaleidoscope, light's reflection engenders new patterns and emergent designs. In this research, the designs are not so beautiful to look at (perhaps it should be called a palioscope, from the Greek word *palios*, or 'not good') but instead they reflect patterns of exclusion and discomfort expressed through conversations and reflections on images of knowing. Included in these conversations are patterns of exclusion due to official terminology, due to gendered messages, due to a world of binary oppositions—all formed as a result of some often unquestioned 'properties' (or ideologies) of knowing. Reflections on each of these patterns for how knowing has been experienced (or, in metaphorical terms, on each of these images created through and by the properties of light) can possibly help to expand beliefs about, and possibilities for, knowing (in)^3 mathematics and science.

Language

In this 'text', light and its properties are used as a metaphor for the explorations of preservice teachers' thinking on what it means to know in mathematics and science. I believe that the properties of light help to illustrate a few parallel properties of knowledge, particularly as experienced by my preservice teacher research participants. School experiences of knowing and learning, described by my participants and myself in the text that follows, bear striking resemblances to some of the properties (or characteristics) of light—where it comes from, how it travels, etc. For instance, in the first chapter of the 'text', I write from the metaphorical perspective of luminous and non-luminous sources of knowledge—comparing knowers who feel capable of constructing, or producing, their own knowledge with those who have really only learned how to receive and reflect back the knowledge of others. From my conversations and my own personal experiences, this reflection analogy aptly describes how many women have experienced knowing in school science and mathematics. I therefore use the metaphor of luminous and non-luminous sources to critically examine how sources of knowledge need to be questioned and cha(lle)nged in terms of what counts as knowledge, who has knowledge, and how knowledge is constructed. In subsequent chapters, I critically explore experiences of transmission models in education and decontextualized learning in math and science through the metaphorical topics of the rectilinear propagation of light and the wave-particle duality theory of light. While mainly focusing on the light metaphor in a critical (and often pejorative) manner, in chapter four the properties of lenses are used as a metaphor to introduce possibilities for other ways of knowing and for other ways of *viewing* knowing in math and science.

One reason for my use of the light metaphor in a critical portrayal of school mathematics and science can be understood through the visual metaphor analysis by Lorraine Code (1991):

> Visual metaphors—knowledge as illumination, knowing as seeing, truth as light—shape hegemonic conceptions of knowledge just as surely as masculine metaphors shape hegemonic conceptions of reason. Vision has so well established a role in mainstream thinking about the nature of knowledge that its aptness is rarely questioned. (p. 140)

There are several reasons why Code (1991) objects so strongly to the privileging of visual metaphors in theories of knowledge. First of all, she writes on how vision and other optical metaphors have been treated as the privileged sense in dominant epistemologies, as if *perception* were solely dependent on this sense. Vision is presented as an "innocent mode of perception, which provides direct, untainted access to reality" (p. 141) and, hence, theories of vision "ignore the extent to which even seeing is culturally, historically, and ideologically shaped" (p. 141). In discussing feminist research that counters the dominant epistemological discourse, Code emphasizes resources which do not privilege vision but recognize that "perception engages all of the senses; objects are known by touch, holding, sensing their whole presence" (p. 148).

Secondly, and perhaps more importantly for feminist epistemologies, Code writes on how vision detaches the viewer from the viewed; the subject from the object. This further de-emphasizes the non-visual senses in knowing (especially the corporeal, since vision is reliant only on the eye) and further emphasizes objectivist ideology. "Whether a philosophical theory is constructed from the vantage point of the 'inner' eye—the eye of the mind—or the outer eye of natural vision, the outcome, in mainstream theory, is some version of the detached (and hence *objective*) relation of knowing subject to objects of knowledge." (Code, 1991, p. 141) This subject-object relation is one I will focus on in the first chapter of this text.

It is with/in this critical 'light' metaphor that my research (re)presentation takes shape. The light metaphor in this text is not intended to support mainstream theories of light as knowing and vision as the only way to know. In fact, I hope that through this metaphor one will appreciate the satirical play with/on the various properties of light in an attempt to critically illuminate preservice teachers' experiences of, and perceptions on, what it means to know (in) mathematics and science.

There is a crack in everything.

That's how the light gets in.

(Cohen, 1998, p. 66)

Another aspect of language in this 'text' that calls for explanation is the reference throughout to the subjects of mathematics (often abbreviated in this text to math) and science combined. It is important to clarify why I chose to study preservice teacher thinking and perceptions of knowing in both mathematics *and* science. The conversations with my participants moved to-and-fro between these subject areas, sometimes mainly referring to one or the other, and sometimes both. The writing reflects this to-and-fro movement— not to suggest that my participants' images/perceptions of mathematics and science knowing are identical, but that the stories weave together to create a kaleidoscopic image of knowing (in) these subjects. The main point being that this research text is not about the *content* of the subjects so much as it is about preservice teachers' perceptions of, and experiences in, *knowing* in these subject areas. They are coupled together because it is the similarities of their epistemological views (that is, what constitutes knowledge and why, according to my participants) that are being explored. Discussions of other school subjects naturally come into the participant conversations also, highlighting comparisons of knowing in different subject areas. In terms of the personal lived experiences and ideas explored in this study, the epistemological discussions of the two subjects are closely related.

Connected to this mathematics and science relationship discussion, it is also worth noting that my research interests first took root in my belief that one of the solutions to the cycle of 'mis-education' in school mathematics and science is to integrate the two subjects in middle and high school curricula. The experiences of my participants (and of others documented in research; see, for example, Briscoe &

Stout, 1996; Ernest, 2004; Mulholland & Wallace, 1996; Roychoudhury, Tippins, & Nichols, 1995; Ball & Bass, 2000) indicate that the loss of interest and perceived ability in math and science learning is most prevalent when the subjects are presented in a decontextualized and abstract manner. My original intention in this research (as described briefly in the Introduction) was to explore mathematics and science integration as a way to address the needs of *all* learners, not just those who are better able to work with such discrete and disjointed concepts. As I learned quickly on the research journey, the 'problem' is much more complex; the issues are more systemic and deeply entrenched than a mere math-science integrated curriculum could ever hope to alleviate. This is how and why I arrived at this critical qualitative study of mathematics and science epistemologies.

Motivation

The motivation for this text emerges form a desire to write and speak with our lives. I believe an illustration of this motivating force is embedded within the introductory narratives. In and through this text that follows, I present voices that articulate the need for making epistemology an explicit agenda of teacher education curriculum. There is a necessity for preservice teachers to reflect on their epistemological beliefs. This may begin by inviting preservice teachers to reflect on their own personal lived experiences as math and science learners in school. And, it is crucial for researchers and teachers to listen as preservice teachers voice their stories and reflect on the meanings these stories have for them (Danielewicz, 2001). Such a reflexive approach is a necessary first step toward actually *living* different ways of teaching, learning and knowing, not just reading about them.

> Uncovering preservice teachers' beliefs about the nature and sources of knowledge could be used as means towards self-reflection on teaching and learning. If preservice teachers become more aware of their "ways of knowing" by observing, discussing or reflecting on them during teacher education, it might stimulate the reexamination of their beliefs about teaching. Studying epistemology as a part of teacher education will also enhance teacher educators' understanding of students' beliefs and assumptions about learning and therefore could be instrumental for designing learning activities that will challenge or contradict some of the embedded beliefs. (Bozin-Mirkovic, 1997, p. 195)

Studying 'ways of knowing' is not, by any means, a new phenomenon. I am often reluctant to even use the expression, as I believe it is always closely associated with the work by Belenky, Clinchy, Goldberger, and Tarule (1986) on women's ways of knowing. So many years later, this piece of research continues to inform and shape research in the areas of epistemology and women's education, including parts of my own research text. One of my criticisms of the Belenky et al. work, however, is that it focuses on the knowers, classifying and categorizing them according to stages of development in their epistemological positions.

Lynda Stone, who conducted a review of Women's Ways of Knowing, is also critical of that research for similar reasons, asserting that a model for educating women should be based on consciousness raising, not on developmental categories. She proposes instead a "feminist trialectic", which moves in three directions to explore relations between self and others: (1) self-reflection on and articulation of one's own beliefs, (2) engaging in group conversations, and (3) critiquing and expanding on sources of knowledge by being critical of experts and authorities while simultaneously valuing one's own theory (Stone, in Bozin-Mirkovic, 1997, p. 42). I believe that my own research design and methodologies embody Stone's description of a feminist trialectic, but with an additional personal perspective and autobiographical flavour to the research. As noted in Winterdyk, Coates & Brodie (2006), "[r]esearch is a social activity, and the processes of designing, analyzing, and interpreting the study are influenced by numerous factors including the author's theory and philosophical position" (p. 1). As the reader will soon realize, motivation and direction for this research emerged from within my own tendency toward self-reflection and articulation of my personal experiences and philosophies.

Turnings

The research text that follows is not just about turning a kaleidoscope to obtain interesting images and stories; I believe this would reflect only a descriptive turn. As Eisner (1998) states, "description deals with what is, interpretation focuses upon why or how" (p. 98). I maintain that it is important not only to *describe* experiences of knowing (in) math and science, but to explicitly reflect on the *meanings* my participants have attached to these experiences. The narrative turn in my research is rooted in the belief that "people make meaning of their lived experiences by narrating those experiences" (Rossman & Rallis, 1998, p. 185).

There is also an important critical turn in my research, in that I am interested in understanding why and how certain meanings have come to be attached to experiences. Carspecken and Apple (1992) write about the importance of dialogic interactions in critical qualitative research, where "the subjects of study are asked to reflect on their lives in ways that may be new to them and to share in the production of a theory relevant to their lives" (p. 513). Apple (2004) emphasizes a critical perspective in writing:

> … we need to examine critically not just 'how a student acquires more knowledge' (the dominant question in our efficiency minded field), but 'why and how particular aspects of the collective culture are presented in school as objective, factual knowledge.' How, *concretely*, may official knowledge represent ideological configurations of the dominant interests in society? (p. 12).

Still turning to theoretical underpinnings, it is important to state at the outset that the social construction of gender is at the centre of this inquiry in that I "see gender as a basic organizing principle which profoundly shapes/mediates the concrete conditions of our lives" (Lather, 1991, p. 71). I believe that alternative discourses (not merely adaptations of current patriarchal ones) must be developed which embrace the belief that "values, politics and knowledge are intrinsically connected" (Usher, 1996, p. 128). Of central importance to feminism is the need to address this complex connection between values, politics, and knowledge. As Weedon (1987) reminds us, "transforming both the social relations of knowledge production and the type of knowledge produced... requires that we tackle the fundamental questions of how and where knowledge is produced and by whom, and of what counts as knowledge" (p. 7). Feminist discourses acknowledge that the personal is political and that each claim to knowledge is historically and contextually located. I echo Lather's (1991) call for the "involvement of research participants in the construction and validation of knowledge" (p. 60) by building reciprocity into the research design, through interactive, dialogic interviews including self-disclosure on the part of the researcher and by negotiating meaning through collaborative theorizing and reflexivity in research.

Along the research journey, through the narratives of the problematic, my participants and I purposefully reflect on what is useful theory/action for their own project in the interpretations of the meanings they create in/for their lived experience. Beyond and within these interpretations, reflection on lived experience catches glimpses of the spaces that could serve our emancipatory interests, exploring the conditions necessary for ideology critique and transformative action. Habermas (1971) indicates that a critical social science is not content with reaching understanding and possible consensus of meaning but that it "is concerned with going beyond... to determine when theoretical statements grasp invariant regularities of social action as such and when they express ideologically frozen relations of dependence that can in principle be transformed" (p. 310).

Turning to methods, the reader might be interested to know that participant observations, individual conversations, and focus group discussions formed the predominant 'data' gathering techniques for the research study. The reason I think *the reader might like to know* this is that such a 'triangulation of data' is a well-documented test of validity (or, as referred to in more postmodern texts, test of *verisimilitude*; see, for example, Roth (2005)) in qualitative research design (Janesick, 1998; Maxwell, 1996). While my methods fit nicely into such a framework, advocated for many genres of qualitative research, I prefer to quote Richardson (1998) in saying:

> In traditionally staged research we valorize "triangulation"... But in postmodernist mixed-genre texts, we do not triangulate; we *crystallize*. We recognize that there are far more than "three sides" from which to approach the world." (pp. 357-358)

The crystallization metaphor, I believe, is a liberating way to view postmodern texts and validity/verisimilitude in qualitative research. Similar to Richardson, my metaphor of a kaleidoscope portrays ideas of "symmetry and substance with an infinite variety of shapes, substances, transmutations, multidimensionalities, and angles of approach" (p. 358), as well as the perspective that reflections with/in a kaleidoscope create "different colors, patterns, arrays, casting off in different directions" (Richardson, 1998, p. 358). As Richardson states, "In postmodernist mixed-genre texts, we have moved from plane geometry to light theory..." (p. 358). And so, in this text, I move to light theory to help visualize the images my participants have of knowing (in) mathematics and science.

> Our choices are compounded by their sheer range – the self we choose to be, our participants and the way we/they present themselves, the forms the texts take, and the audiences with whom we and our research participants wish to interact via the text. (Lincoln, 1997, p. 39)

Format

As can be seen by quickly scanning the pages of this text, it has not been constructed according to a traditional linear approach to (re)presenting academic research. In fact, I frequently refer to the research text as a textextextext— where the repetition (and changing font size) of the word *text* extends and re/informs the possible meanings and multiple layers embedded within.

> In a postmodern world, where interpretation and translation are paramount, we – as subjects, objects, authors, and narrators – will often feel lost and ill at ease. The point, of course, is not to ignore the discomfort or to wish it away, but to confront the issues of identity and representation and consider how we might develop texts that highlight the problematic worlds we study, our relationship to such worlds, and how we translate them. (Tierney, 1997, p. 34)

In this section, I will discuss several key aspects of the textextextext which relate to its format—including why *this* format and not a standard linear format. Firstly, I would like to introduce the ideas of writing-as-research textextextext. In keeping with the multiple layers of my research explorations into what it means to 'know,' I am informed by Laurel Richardson's (1998) perspective on writing as a way of knowing. She discusses the importance of the writer writing herself and her voice(s) into the research— to write as a method of discovery and inquiry, not simply "writing up" the completed research product. Richardson (1992) reflects on the routine process in which an interviewer produces interview text,

> … as though information consists of facts or themes or notions which exist independently of the context of facts or themes or notions which exist independently of the context in which they were told, as if the story we have recorded, transcribed, edited, and rewritten as snippets is the true one: a 'science story' (p. 23).

I am interested in the unfolding nature of the research story, in that I am writing about the happenings, strategies, and motives of the characters as *the* textual practice. In other words, I attempt to create a form of research (as) writing (as) research that highlights the unfolding stories of struggles, questions, and rewards, presented as *the* word not the after-word. Clandinin and Connelly (2000) support this approach, writing that "… those engaged in narrative inquiry know that the written document, the research text, like life, is a continual unfolding in which the narrative insights of today are the chronological events of tomorrow" (p. 166).

Secondly, I seek to highlight the importance of creating a reflexive and polyphonic textextextext. In recent qualitative research endeavours, much attention is paid to the issue of voice—the reflexive voice(s) and the polyphonic chorus that can be (per)formed through/in textual practice. As Lincoln (1997) proposes, "[m]ultiple stories feed into any text; but, equally important, multiple selves feed into the writing or performance of a text, and multiple audiences find themselves connecting with the stories which are told" (p. 38).

Lincoln (1997) asks the question: "What are the implications for this polyphonic chorus of author/selves, subjects and participants, audiences, and texts?" (p. 38). I think one very important implication is the potential for a rich, multi-voiced tex-textextext that does not attempt to speak unequiVOCALly for all. What/who are the storied voices emerging in my polyphonic textextextext? The multiple layers acknowledged in this textextextext include the voiced re-presentation *text* (author, participants, responsive colleagues, and scholars in the field) with/in the performance *text* (poetry, prose, comedy, journal), with/in the metaphorical *text* (properties of light, diagrams, photos, vocabulary).

And finally, as mentioned in the overview, the format has been constructed to resemble that of a physical science textbook. I use metaphorical language in the textual performance of this physical science/research textextextext. The metaphorical language of the physical science topic of light weaves throughout my writing, as a way to both underscore the influences of textual authority and also the nature of language in mathematics and science. Such light-based metaphors are present in our everyday experiences and help to illustrate many of the experiences of knowing voiced by my participants. This form of textual play and performative writing has the potential to appeal to a more diverse audience of readers, acknowledging, however, that one text cannot be written for everyone.

> Poststructuralism suggests two important things to qualitative writers: First, it directs us to understand ourselves reflexively as persons writing from particular positions at specific times; and second, it frees us from trying to write a single text in which everything is said to everyone. (Richardson, 1998, p. 349)

Clandinin and Connelly (2000) encourage novice narrative inquirers to engage in an ongoing experiment with narrative form. Thus, narrative inquiry is not simply a retelling of stories but it involves decisions about *what* is retold and *how* it is retold. The unique form of this text is not just a random or experimental event; the textextextext has been written based on *how* I think the complex and continually reflexive nature of my research topic is best re-presented, just as that which has been re-presented in the stories themselves (the 'content') is selected to reflect *what* I think is most significant for people to know about. It is difficult for the reader to know, in terms of stories and content, *what* I have left out. The format, however, is much more open to criticism, since traditional guidelines for *how* to write narrative inquiry and academic research abound. The reader might ask herself if this format is 'better' than a traditional, linear representation, but I believe this question is not the best one to ask. Such a question likely emerges from within a reader's well-established writing/reading comfort zone. In other words, a reader might resist being nudged into thinking differently while being interrupted along the journey by the multiple layers and directions of thought. It is crucial to remember, however, that the topic of my research is typically not a comfortable one for elementary preservice teachers; it is one that has jagged and unfinished edges. At times, the conversations jab and stab my participants into anger and frustration while, at other times, the conversations provide them with a sense of relief from trying to convince themselves that things are smooth and comfortable in the world of preparing to be a math and science teacher. The metaphor of a kaleidoscope would not have been chosen if the patterns of conversations were predictable and linear. Within this text, there are many different and unique patterns to be seen if one takes the time and energy to continually turn things over in her/his mind, watching the colourful bits and pieces slide about and arrange themselves into images. The colourful bits and pieces are arranged in this text in the form of reflexive moments, poetic sojourns, research insights, critical segues, etc. There is no order in which to view them, but they all do participate in the overall precarious designs that emerge.

With this format, I am asking the reader "to muster a certain tolerance for the unease that may accompany ambiguity" (Clandinin & Connelly, 2000, p. 154), since the format of the text represents and reflects some of the ambiguities and tensions of both the research topic and the research (as) writing. As I have written elsewhere (Nolan, 2005), the kaleidoscopic writing format for my research text encourages me (and hopefully a few readers as well) to imag(in)e postmodern possibilities for mathematics and science education. The kaleidoscopic textextextext is a way of embodying different ways of knowing (in) a research/academic text which is, in turn, about embodying different ways of knowing (in) mathematics and science.

> My desire [is] to not only write the research text *about* different ways of learning and knowing (in) mathematics and science but to actually write the research text *through* different ways of learning and knowing. The meaning behind the prepositional choice of 'through' over 'about' highlights the intimate connections between research, learning, and knowledge, proposing a paradigm shift for understanding and realizing the possibilities of a performative text. (Nolan, 2005, p. 120)

Features

The (re)presentation format that I have described helps to portray the layered nature of this qualitative research study. The performance textextextext includes researcher and participant reflexive voices, response data, autobiographical snippets— all viewed through the lens(es) of metaphorical narrative, woven into a physical science textbook. Several approaches are taken to perFORM the text.

As mentioned previously, the special features of the text have been chosen to reflect the reflexive and multi-faceted nature of the research, as well as the ambiguities and tensions experienced by the research participants. Along the textual journey, the reader will experience linear-reading interruptions in the form of metaphorical links (**`Did you know that…?`**), related research connections (**`In-`**

side Research), researcher reflexive moments (**Inside (my) Research** and **Reflexions**), response data from colleagues (**RESPONSibilities**), and related ponderings (**A Matter of Perception** and **Focus on...**). Also, in an attempt to capture the contexts of the various conversations, extended transcript excerpts are re-presented in the form of newsletters and "**conversations with...**" fragments. When reading these various kaleidoscopic features, it is worthwhile to note the temporal weavings with/in the text since they can provide glimpses into the emergent and unfolding nature of the research journey. For instance, the response data from colleagues (**RESPONSibilities**) reflects musings and thoughts on the text as it was developing and evolving (as it is *still* developing and evolving?). One motive for highlighting these voices is to provide a context for some of the discussions and questions initiated by readers.

I would be amiss if I did not highlight the features of this textextextext which pertain particularly to my own personal lived experiences of writing (as) research. While my learning experiences are well-grounded in positivistic epistemology and scientific method, I am attempting to honour my desires to break free from some of the associated constraints, just as my participants and I discuss such imaginative and creative departures in the teaching and learning of mathematics and science. To do this, I have chosen to embody poetic and visual forms of representation throughout the text. Richardson (1998) proposes that "poetic representation reveals the process of self-construction, the reflexive basis of self-knowledge, the inconsistencies and contradictions of a life spoken of as a meaningful whole" (p. 25). The poetic form of representation is an embodiment of lived experience, where "[k]nowledge is thus metaphored and experienced as prismatic, partial, and positional, rather than singular, total, and univocal" (Richardson, 1992, p. 25). My poetic and visual sojourns are sprinkled throughout the text, and I shall begin the journey through this physical science textbook with one such piece, written while on a research (as) writing retreat...

Knowledge redefines and extends the boundaries of what we don't know.
This is why knowledge often feels like loss rather than acquisition.
(Michaels, 1995, p. 180)

conSCIENCEness Land(e)scapes

translation: escaping to lands of being with science

the division between
my body, my mind is never so sharp
moving away from the computer
   to hear the SNAP
   SyNAPses ceasing
   joints reminding
my mind, my body
I've not stood for hours
and where am I after all?
   the minutes tick by
   the pages
      unfinished links
      [insert here]
      by the asterisks ***
      outside the box
      weaving a web of colour
   and on the porch
   the tangled fly reminding
my body, my mind
I've not eaten for hours
and where am I after all?
   on the landing escaping
      laptop hum
      unending references
      unfinished busyness
   the power surges
         and my back is to the wall reminding
my mind, my body
mind and body
[caught in dichot...]
o my
bodymind
[SNAP]
and where am I after all?

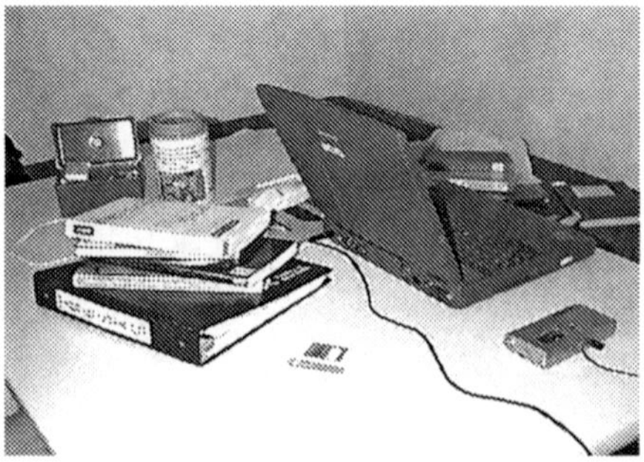

Where does light come from?

Knowledge does not arrive by studied labour;
though you hunch year on year above your books
and Bachelor, Master, Doctor of each science,
labour farther to enlarge your mind,
you cannot come to knowledge;
that eludes deliberate effort as does ecstasy,
and must happen suddenly,
your startled understanding falling to its knees.
(Skelton, 1998, pp. 214-215)

WHERE DOES LIGHT COME FROM?

preVIEW...

In this chapter, luminous and non-luminous sources of light are discussed as an introduction to the notion of 'sources of knowledge.' Upon closer examination of the luminous/non-luminous dichotomy, several additional binary oppositions are revealed and deconstructed, in an attempt to understand inherent tensions and complexities. Western epistemological views perpetuate the dichotomous tension of choosing one side OR the other—'good at' or 'not good at'—creating no possibility for residing in the in-between spaces. This chapter explores the existing problematic nature of such oppositions as well as see(k)ing possibilities for discursive movement in the 'slash' space of binary representations. The discussion sets the stage for explorations and ponderings into these in-between ambiguous spaces as they relate to preservice teachers' images of what it means to know (in) mathematics and science.

Most objects in our world—

—do not *produce* visible light. We see them only because they *reflect* light from

These 'producers' of light are called **luminous** sources of light—some are natural and some artificial, but *all* emit their own light energy and that makes them very illuminating indeed!

Rose (1994) writes on the issue of the under-representation of women in the sciences, offering two possible explanations which, she asserts, emerge from essentialist positions. She states that either the 'problem' of under-representation lies in the nature of women themselves OR it lies with the exclusionary processes embedded in educational systems. (Sounds like a positivistic, dichotomous view, doesn't it?) Based on interpretations of the history of science, where those who typically "do" science are Western, middle class, white males, it may be relatively straight forward to understand why the focus for explaining under-representation of women is more commonly on the former than the latter essentialist claim.

In writing about a feminist successor science, Harding (1992) states that "the Enlightenment vision explicitly denied that women possessed the reason and powers of dispassionate, objective observation required by scientific thinking" (p. 345). Through Enlightenment idea(l)s of searching for *the* Truth, "for the ready-made world that is out there" (Harding, 1992, p. 346), characteristically dichotomous views have been cultivated in Western thinking. The Enlightenment idea(l)s are described by Bernstein (1976) as the belief that "knowledge, and in particular detailed empirical knowledge of how society and politics really work, is supposed to provide the basis for enlightened action and social reform" (p. 53). Keller (1995), in pointing out important omissions in the social study of science, argues that simply rejecting the idea of "scientific neutrality" is not going far enough. She states that the social study of science still pursues "its critique in terms that tacitly support the divisions between public and private, impersonal and personal, and masculine and feminine: divisions that continue to secure the autonomy of science" (p. 7). It is the problematic nature (and noncritical preservation) of these false dichotomies that is the subject of this chapter.

Speaking on binary oppositions, Joan Scott (1997) states:

> Fixed oppositions conceal the extent to which things presented as oppositional are, in fact, interdependent—that is they derive their meaning from a particularly established contrast rather than from some inherent or pure antithesis.... If binary oppositions provide insight into the way meaning is constructed... then analysis of meaning cannot take binary oppositions at face value but rather must "deconstruct" them for the processes they embody. (p. 761)

Deconstruction should not be associated with destruction. Instead, with/in this dichotomous context, deconstruction is best viewed as an attempt to expose the subtle meanings that are already deeply embedded in the socially constructed (and hegemonic) binary oppositions. Dimitriadis & Kamberelis (2006) offer a valuable perspective on the potential for deconstruction as a way of reading:

> "... deconstruction is a form of what has long been called a *critique*. A critique of any theoretical system is not an examination of its flaws or imperfections. It is not a set of criticisms designed to make the system better. It is an analysis that focuses on the grounds of that system's possibilities. The critique reads backwards from what seems natural, obvious, self-evident, or universal, in order to show that these things have their history, their reasons for being the way they are, their effects on what follows from them, and that the starting point is not a (natural) given but a (cultural) construct, usually blind to itself. " (Derrida, 1972, p. xv)

> It is in the process of reading closely, with an eye for how an idea is constructed, that one also comes to see the points of potential rupture, the cracks and other points of instability within the structure. It is in the process of close reading that one sees deconstruction happening. (p. 104)

By delving into the structure of binary constructions, one underscores the interdependence of the terms and exposes the ruptures or cracks within. Might I suggest that we let the dichotomous deconstruction begin with…

good at / not good at

right / wrong

subject / object

A Moment in HERstory

My experiences of tutoring girls in grades 8-10 mathematics provide a main motivating force for me in my research. The first the girls often said to me was that they were not good at math. They clearly perceived people to be either 'good at it' or 'not good at it,' nothing in between. When I taught mathematics to preservice teachers at a teachers' college, I used to ask some of my students (who seemed somewhat disinterested in, and occasionally frustrated by, mathematics) how they came to choose this subject as their middle years teaching minor. They would inform me that the draw to mathematics for them was predominantly because of more job prospects. This rationale troubled me. It reinforced my fear that a cycle of mathematics miseducation would continue. Ball (1990) proposes two categories of students in teacher education programs: the 'haves' and the 'have nots' in mathematics. Neither category of knowledge, she claims, is desirable for the teaching and learning of mathematics because students of both categories tend to reinforce the procedural learning they experienced, only now in the role of *teachers* of mathematics. In my opinion, neither category is acceptable as a way to *describe knowledge.* While Ball provides potentially informative research on the cycles of "miseducation" and "dispedagogy", I believe that the continued use of these categories signifies that we are hopelessly caught in dichotomy. On which side do you (re)side: have **/** have not? *How should you know?*

In this chapter, I make connections between the metaphorical dichotomy of luminous/non-luminous sources of light and the general notion of dichotomies in Western epistemology. As a western society, we are constantly categorizing: the one who knows and the one who doesn't know, the one who has and the one who doesn't have. We thrive on these dichotomies. If knowledge, however, is portrayed as either/or, then the space between (the slash space) is forever neglected. This seems to be the most hegemonic aspect of schooling, in my mind. It creates and expands the gap between mathematicians and scientists (those who 'know' (in) the field) and the public (those who don't 'know'). In light of this, the gap between theory and practice, between knowledge and its applications, is ever widening (but with nothing happening in this gap/space except for the creation of an “intellectual and practical vacuum” (Bernstein, 1976, p. 53).)

Before discussing possibilities for imagining positions and spaces other than the extremes of these oppositional binaries, I will explore how these standard notions of dichotomies originated and how they tend to manifest themselves in ways detrimental to learners. To do this I will first look at the dichotomy of right and wrong (answers): a dichotomous tension that is explored in great depth throughout this text.

Ask the Book Lady

Daughter can count on the future

Dear Book Lady: My daughter is just beginning Grade 4 and already she says, "I'm no good at math." Can you help?

Critically acclaimed author/editor Valerie Wyatt wrote her newest book, *The Math Book for Girls and Other Beings Who Count*, just for girls like your daughter. Wyatt knew that 1990s statistics show that women represented only about 10 per cent of the scien-

book's) illustrator Pat Cupples show how mathematics is useful to her a artist. "Nora isn't hard to draw," Pat, "as long as you keep mathem in mind."

Read together the interviews women about math and jobs. Y to lear ter to

("Daughter can count", 2000)

> The Enlightenment taught Western peoples a way of thinking based on logic and reason. The great thinkers of the period laid the foundation for the modern age of rationality, which left no doubt as to the right and wrong answers to any question. (Kincheloe, Steinberg, & Tippins, 1999, p. 129)

In striving to achieve Enlightenment goals, learners were/are covertly taught to place themselves in little essentialist boxes that serve to label them and their capabilities. They learn to play the necessary learning game in order to create the least amount of turmoil.

> Soon after children enter the halls of the public school, they quickly learn how to answer questions without enquiring further. "Playing the game" has become a pastime in the schools, the workplace, and even the home. A "smart" child knows when to be quiet and how to answer appropriately. Children who have difficulty with schooling many times do not "know" the rules of the learning game. Successful children are often those who must become secret knowers, who view the world of learning from a protected shell that allows them to answer publicly with the appropriate response and to know privately that something is seriously wrong with how they are being educated. (Kincheloe et al., 1999, p. 129)

[Coming Up Next: *In My Very Own Living Room*]

In my very own living room.

I used to love watching Miss Anne. She was famous in my eyes, with her daily show televised in my very own living room. And all those lucky kids who got to sit in a circle with Miss Anne, telling stories, playing games, and journeying to foreign lands across seas in their imaginations.

*In **all** of our imaginations. I didn't just watch the show. As my mother and older sister would attest to, I participated. And I learned. When Miss Anne and all the girls and boys ventured to foreign lands, trying to make the journey by car, or train, or boat, **I** knew the end of the story. They were going to have to travel by airplane. They did yesterday. And the day before. And they would tomorrow, too. It was just something I knew after so many travels with Miss Anne. Even at the age of four.*

Anyway, this gives you, I'm sure, a picture of the passionate involvement and learning that took place in my very own living room. So you can imagine my excitement when I was chosen to **BE ON THE SHOW!** *I don't remember exactly how it happened but, in retrospect, I'm pretty sure now that selecting me had nothing to do with how smart I was.*

*Probably because of my tremendous excitement (and maybe also because of many years now having come to pass), I am pretty foggy about **that** entire episode of Miss Anne. Except for two things. My knee socks kept falling down and required a lot of my concentration. I felt the need to look respectable on (semi) provincial TV. The last thing I wanted was for other kids, in **their** very own living rooms, to point and laugh at the girl on TV with one sock up and one sock down. It happens you know.*

I remember one other thing. However much I was concentrating on my socks, I still remember being excited about the upcoming *imaginary trip to some foreign land. It didn't matter where we went really. I was ready. I could hardly wait for the time when we would all stand up and "fly" through the room as airplanes, on our way to an exotic faraway destination. I sat eager with my answer to Miss Anne's question of "well girls and boys, how will we get there?" Finally, she asked it and I could contain myself no longer— "AIRPLANE!" I volunteered excitedly. Much to my chagrin and dismay, Miss Anne had heard not my response but another child's "car!" I tried again. **"AIRPLANE!"** How was it that she heard "train!" from that little boy and not my **"AIRPLANE!"**? Hmmm, I thought. A little louder, looking directly at her, and I'm sure she'll hear me this time. **"AIRPLANE!"***

*Very carefully, while still focussed on the little girl waiting for approval on "boat!", Miss Anne leaned over to me and whispered "Shhhhhh. Not yet. I'll get to **you** later!"*

At only four, I learned some very important things.

There is a time for everything.
A time to be wrong and a time to be right.
A time to wait and a time to push forward.
A time to be heard and a time to sit silently.
A time to learn new things and a time to pretend I didn't.

A time to move with the flow of the planned lesson, which is also the time to not feel free to express my ideas or construct my own context for learning.

A time for being a public learner and a time for learning only in my very own living room.

These are the best of times. These are the worst of times.
If only I had the wisdom, at four, to know the difference.

In an earlier quote about learning to play the game (by Kincheloe, et al.), several dichotomies come to the forefront: right/wrong, successful/unsuccessful, public/private, subjective/objective. And, as with most dichotomies, there is an implicit privileging of one element (the first) over the other (the second).

> In mainstream theory of knowledge, a persistent set of dichotomies sustains a (possibly unconscious, often implicit) institutionalization of 'masculine' modes and a concomitant denigration of 'feminine' modes in philosophical styles, problems, and methods. Feminist theorists have argued, persuasively, that dichotomous thinking is peculiarly characteristic of malestream thought.... But the idea that there are connections between these dichotomies and male intellectual hegemony is prima facie plausible just because of the fact that so many deeply entrenched dichotomies run parallel to a taken-for-granted male/female dichotomy, not just descriptively but evaluatively. (Code, 1991, p. 28)

What I hope to highlight in this chapter (and in this text in general) are possibilities for challenging the exclusionary principles involved in dichotomous thinking and, in doing so, create spaces for questioning some very basic epistemological assumptions concerning sources of knowledge. In terms of the metaphorical dichotomy of luminous/non-luminous, I question if it is possible to acknowledge **both** luminous **and** non-luminous sources in knowing, as well as the ambiguous spaces in between. My use of this metaphor is not an attempt to reify the opposition of the two words but to actually work to dissolve the hierarchical division between the two by proposing a view of *knowledge as interplay between the two*. Code (1991) reminds me that "the dissolution of a dichotomy does not render its terms meaningless. Rather, it denies both terms the absolute force that the oppositional structure of the dichotomy confers" (p. 30). The hierarchical and oppositional nature of dichotomies establishes both oppressive constraints and unrealistic relationships between/among the 'elements'. I want to dwell with/in this metaphorical dichotomy as a way of highlighting how all knowers are both luminous and non-luminous sources of knowledge.

> ***Bi: two—only??***
> ***why not all the twos [tous les deux]***
> ***and the between/s.***
> (Cixous & Calle-Gruber, 1997, p. 51)

There's always some reason to feel
not good enough.
And it's hard at the end of the day.
(McLachlan, 1993, track 4)

not to erase
not to reverse
not to replace
                                        maybe to displace
                                        or even disperse
to cross and recross
to pluralize and multiplize
to slur and blur   the boundary b/w the two

"knowledge is in the end based on acknowledgement"

(Wittgenstein, as quoted in Code, 1991, p. 215)

> Yet the capacity to gain acknowledgement is gender-related in ways that Wittgenstein himself would not have envisaged, and its gender-specificity is multidimensional. Like any other knower, a female knowledge claimant has to claim acknowledgement from other participants in a form of life. But advancing such claims is as much a political action as it is a straightforwardly epistemological one. Before she can so much as seek acknowledgement, a woman has to free herself from stereotyped conceptions of her 'underclass' epistemic status, her cognitive incapacity, and her ever-threatening irrationality... Politically, spaces have to be created where a woman's knowledge can be judged sufficiently authoritative to deserve acknowledgement, and the spaces have to be constructed variously, to respond to differences between and among women. The creation of such spaces depends on the collective and mutually enabling efforts of women who can trust themselves to know, and to know that they know. (Code, 1991, pp. 215-216)

To acknowledge oneself as a luminous source of knowledge is not an easy task. Historical threads of oppression, politics, and power weave together to create a complex fabric of epistemic dis-empowerment. We are taught to rely on authority— on the 'experts'— to verify and legitimize our knowledge. Belief that objects of knowledge are separate and independent from knowers reinforces the flawed premise that knowledge is a fixed entity, 'out there' waiting to be discovered. This theme often emerges in conversations with my participants and it surfaces in many different voices.

conversation with Sylvia and Harriet

Will the experiment come out 'right'?

Sylvia: we tried it [the science mini-experiment] three times... and then said, 'oh, I wish we had [the teacher's] home number; *he'll* know how to do it, and he can tell us and then we'll do it *right...* 'cause we CAN'T figure out how to do it right, we CAN'T do it

Kathy: so the idea of an answer that was 'up there' and you knew it was there... almost like it was just trying to make you mad?

Sylvia, Harriet: exactly (laugh)

Harriet: it was not a gratifying moment

Harriet: this makes me wonder *why*; because with other things, I'll try it until it comes out or I'm not always looking to myself for *the answer*. But I don't know if that's because... it's that mystique of science and how I had always been *taught* science: the teacher knows the answer. There's this one *grand* thing

Sylvia: this science class, I think, helped us experiment to begin with because we became more comfortable with [the teacher] and together as a class, with experimenting. So we were willing to do that, but then when we didn't get the right answer, we immediately retreated...'oh, we can't do it...'

Harriet: immediately, yeah, 'we can't do proper science'

With science, they never asked you 'what do you think...?'
or 'how could you...?'
Science was always 'fill in this blank space here', right or wrong,
half a mark here maybe, if you're lucky
(Helen, December 1999)

You can't possibly know it all.
It's always changing.
Know the process:
know how to find out the information,
and what you need to find out the information,
and then get back to whoever is asking you the question.

It's written down *somewhere*, *someone's* documented the answers, *someone* has access to the information. You just have to know how to get it.
You can't know it all.
(Maude, December 1999)

That's what all those tests were...
it was like: 'you know it,
you know what I want you to know, or you don't'
and it was an X or a check mark
there was no discussion
it was just a right or a wrong thing
And I guess that, to some extent, science is probably a right or a wrong thing
but I think there could have been some discussion

(Ursula, October 1999)

Question Corner...

Why would someone say they weren't good at science?
(Kathy's question)

Probably because they didn't get somebody's right answer.
(Ursula's answer)

The lived experience of these research participants, as glimpsed in these quotes, suggests that their learning of science was void of the personal. I do not mean that it is necessary to claim epistemic empowerment to the extent that one rejects any and all reliance on 'expertise'. To do this, would be to turn inward so that one occupies only a "subjective epistemological position" (Belenky et al., 1986). It is necessary, however, to gain a more flexible perspective on 'expertise.' Code (1991) states that expertise is not "a neutral entity, part of the 'nature of things'; it is constituted and impeded by social structures, mediated by power relations, and open to deconstruction and revision" (p. 220). From my conversations, it became quite clear that reliance on expertise was much more prevalent in the subjects of math and science than in any other school subject. My research participants gave clear indications that expertise and authority were crucial to them in order to have their answers validated and, thereby, determine if they were good at math and science. As I came to know these women through our conversations, however, it was evident that their epistemological positions varied dramatically between school subjects and personal experiences. Perhaps this is why *Women's Ways of Knowing* became heavily critiqued for essentializing women and placing them into particular epistemological boxes (Belenky et al., 1986, to be discussed further in chapter 3). My participants were clearly not "received knowers" very often outside of math and science, and even when they were, they were usually much more critical of 'the system' (for making them that way) than of themselves. For instance, these preservice teachers clearly expressed their beliefs that a language, such as French, could not be effectively taught and learned word by word, out of context and without practice. So, if they could not "do" French after a dry class of memorization and repeated verb conjugations, it made perfect sense to them to be critical of the pedagogical process or the curriculum or *something else external to themselves*. Yet, they were inclined to interpret similar (unsuccessful and meaningless) experiences of writing and repeating definitions in science as their own personal weakness or failure.

RESPONSibilities

Student views of science & knowledge… we wonder how it can come out so different than we hope.

In research that has been done with girls at various stages of self-development, interesting tensions are revealed in terms of the complex dynamics at play between experienced ways of knowing. Debold, Tolman, & Brown (1996) express concern that the development of epistemological categories further reifies the division between mind and body, making self-authorization increasingly difficult as girls move through adolescence.

> Within a Western culture that authorizes as knowledge the products of a mind abstracted from material reality— of the body, of human relationship, of the particulars of people's lives— we have found that girls coming of age and coming to know the dominant culture typically find themselves torn and, ultimately, split from their own power to authorize their experiences as real and as knowledge. (Debold et al., 1996, p. 86)

We're taught to use our experience as knowledge
and yet it's disqualified if it doesn't fit in.
Why?
(Roberta, December 1999)

Perhaps an acknowledgement of a need for some discernment (or balance) regarding expertise is one thing I secretly desired would emerge with my participants, as we conversed about the importance of answers, the influences of their teachers (and others in positions of authority), and their perceptions on the nature of knowing (in) math and science. I consider these issues to be **at the heart of** my research desires.

verb, transitive
heart·ed, heart·ing, hearts

Archaic.
To encourage; hearten.

— idiom.

at heart

In one's deepest feelings; fundamentally.

by heart

Learned by rote; memorized word for word.

do (one's) heart good

To lift one's spirits; make one happy.

from the bottom of (one's) heart or **depths of (one's) heart**

With the deepest appreciation; most sincerely.

have (one's) heart in (one's) mouth

To be extremely frightened or anxious.

have (one's) heart in the right place

To be well-intentioned.

(The American Heritage Dictionary of the English Language [AHDEL], 1992)

word play...

To have my ***heart in my mouth*** does not ***do my heart good*** and, in the ***depths of my heart***, I would rather have science ***at heart*** than know science ***by heart***. I sure hope my ***heart's in the right place*** in this research.

Discernment of external sources (of knowledge) and acknowledgement of internal sources (of knowledge) is a balance that is not easily achieved. The looming authority of the experts tends to oppress and undermine women in how they view their own personal capabilities as knowers. This is how/why I believe the notion of dichotomies is so influential in this research. Women, who are struggling to gain their own epistemic position of power and authority, will often judge themselves and their cognitive capacities harshly. They may often quickly dismiss their own ideas, mistakenly replacing them with those of external authorities. It is often an either/or situation. Either this is right, or it is wrong. Either I am good at it, or I am not good at it. Either I know (it), or I do not know (it). There is seldom recognition of the personal participation of the knower in the acts of knowing which would acknowledge and legitimize partial knowing.

What needs to be remembered is that "intricate structures of power are at work throughout any social order to legitimate knowledge and the institutional practices it informs"
(Code, 1991, p. 201)

A Moment in HERstory...

At a recent interview for a faculty position, I gave a brief overview of my research. I presented my ideas about the need for preservice elementary teachers to feel less dichotomous tension in deciding on, and then announcing, whether they were 'good at' or 'not good at' math and/or science. I stated that, as teacher educators, it is important to encourage and legitimize other ways of knowing and partial knowledge with our elementary teachers-to-be. What is the point, I asked, in isolating them even further from math and science than they already are after their own school experiences?

One of the audience members (*who was also one of three on the selection committee*) made a comment during the question and discussion period. HE said,
"I just wonder what the parents would say if we told them that their children's teachers had only *partial* knowledge."

His sarcasm got my attention.
His attitude as a teacher educator troubled me.
And his position of power further strengthened my point.

(*By the way, I was not offered that faculty position.*)

Maybe we have been 'telling' the children's parents for too long now that their children's teachers know all, with nothing partial about it (including the non-partial pressure teachers must feel!). When a person becomes a teacher, how is it that they can be expected to somehow magically become a luminous source of knowledge? As students, we were trained to "see" our teachers as luminous sources of knowledge and all we did, as non-luminous sources, was reflect back the knowledge 'given' to us. Then suddenly, when placed in the same classroom as a teacher, we are expected to be 'sources' of knowledge. This view would be very acceptable to me if we could view everyone as luminous sources of knowledge—that students are presented with opportunities to construct their own knowledge, luminous and legitimate.

lu·mi·nous (l¡'me-nes) *adjective*

1. Emitting light, especially emitting self-generated light.
2. Full of light; illuminated. See synonyms at bright.
3. **a.** Easily comprehended; clear: *luminous prose.* **b.** Enlightened and intelligent; inspiring: *luminous ideas.*

[Middle English, from Old French *lumineux*, from Latin *lúminosus*, from *lúmen*, *lúmin-*, light.]

knowl·edge (nŏl′ĭj) *noun*

1. The state or fact of knowing.
2. Familiarity, awareness, or understanding gained through experience or study.
3. The sum or range of what has been perceived, discovered, or learned.
4. Learning; erudition: *teachers of great knowledge.*
5. Specific information about something.
6. Carnal knowledge.

[Middle English *knowlech* : *knowen*, to know. See know + *-leche*, n. suff.]
(AHDEL, 1992)

word play...

know·lous (*adjective*)
1. emitting *knowledge*, especially emitting self-generated *knowledge*
2. full of *knowledge*

RESPONSibilities

Does the idea of *knowlous* describe what we think about learning? Is *knowl* emitted?

Are you a **knowlous** source? I think we all are. Though I do enjoy playing with language (and inventing my own adjectives), I want to clarify that I am not trying to represent knowledge as a commodity to have and to hold— though one often speaks of knowledge in this way— but I *am* trying to emphasis that knowledge is self-generated (self-constructed). This is the point that is often (a)miss(ing).

Her knowledge
His knowledge
A possession
A commodity
On the edge of knowing — almost there
On the ledge of knowing — feel the edge
walk the ledge

(k) now led (ge) toward what?
now led astray

The Enlightenment ideal(s) of the objects of knowledge as inert and unaffected by the knowing processes does not leave any space for interaction and personal participation of the knower. I am drawn into posing several perplexing questions…

How can other sources of knowledge be seriously recognized and legitimized
if we continue to call them 'other'?

How can the dominant discourse(s) be scrutinized and deconstructed
if we continue to reify the hegemony of one discourse dominating another?

How can the alleged gender neutrality of knowledge, culture, language, and history be accepted
if we are scoffed at for being unnecessarily feminist in writing herstory or womankind?

How can women be subjects of/in the acts of knowing
if we continue to make women as subjects the objects of our studies?

In writing about the history of relations of men's dominance over women, especially in terms of what and whose experiences become socially constructed as knowledge, Dorothy Smith (1991) explains the meaning and significance of patriarchy:

> It is as if the world is thought from the position of a consciousness which has its centre in a ruling class of men… From this centre women appear as objects. In relation to men of these classes women's consciousness does not appear as an autonomous origin of knowledge, an authoritative perspective on the world from a different position in and experience of it. Women do not appear to men as men do to one another, as persons who might share in the common construction of a social reality. (p. 250)

Not only do women not appear to men as men appear to one another but women do not appear to *women* in ways that are necessary to transform notions of who has authority and knowledge in our society. (Read that sentence one more time! ☺). Women can have difficulty asserting their own authority as well as experience difficulty accepting the authority of other women's voices. But what else could be expected? After all, as a society, we have been well trained to strive for certain standards, without fully reflecting on and understanding that the standards to which we aspire have been set by men, based on men's experiences.

RESPONSibilities

> So, are you suggesting that women <u>must</u> write differently both in content and style to uphold their own identity and to be recognized? Must we continue to *whine* about being women? Why can't we just move along? Press forward? There is nothing to be gained in over-analysis of what was; it is more critical to determine what might be or what <u>will</u> be! Or what <u>can</u> be!

As women, "we can assert that there is not one way of seeing the world, not one way from which it may be known. There is not one universal subject from whose perspective knowledge can be simply transformed into an objective and universal account" (Smith, 1991, p. 252).

As can be seen by the discussion thus far, one pervasive dichotomy is that of subject/object. It is one that feminist theorists in particular have examined consistently and carefully in recent years. I would like to spend some time pondering this dichotomy. In order to understand how subject and object have been dichotomized, it is worthwhile to first spend some time understanding notions of objectivity and objectivism.

> "The ideological practices of our society provide us with forms of thought and knowledge which constrain us to treat ourselves as objects." (Smith, 1991, p. 251)

Bernstein (1976) defines objectivity to mean "that in any domain of human inquiry—whether physical phenomena, or an existing political system, or even the interpretation of a text—there are intersubjective standards of rationality or norms of inquiry by which we attempt to distinguish personal bias, superstition, or false beliefs from objective claims…" (p. 111). Bernstein clearly distinguishes objectivity from objectivism, the latter of which he states "is a substantive orientation that believes that in the final analysis there is a realm of basic, uninterpreted, hard facts that serves as the foundation for all empirical knowledge" (p. 111). Objectivism is an orientation that coincides with the "myth of the given" and it is not, as some might like to believe, "an innocent mistaken epistemological doctrine" (p. 112). "It has dangerous consequences insofar as it tends to distort and reify 'facts' which are historically conditioned—'facts' which reveal only one among the many different possibilities that human action may take." (p. 112)

In exploring why such a chasm occurs between objective and subjective knowledges, Bernstein (1976) writes about the work of Charles Taylor, stating that he

> … challenges the categorical distinction that runs so deep in mainstream social science: that phenomena must be either classified as objective—"out

> there", and consequently in the category of the observable— or lumped together as beliefs, attitudes, and opinions which are "merely" subjective and private. (p. 113)

Such categorical distinctions do not run deep only in the social sciences. The Western view of science is characteristically dichotomous and it runs ever so deep into school curricula.

Quote of the day...

the teacher knows the answer... there's this one grand thing
(Harriet, December 1999)

If there is "this one grand thing", in the way of a single answer, then no room for ambiguity exists in this dichotomous picture.

> ... feminist theory does not fear that the act of learning can include, and even be intimate with, ambiguity. Patriarchal or modernist ways of knowing restrict the learner to the fragmentation of information and an insistence on determining the answer. Ambiguity is inconceivable, imagination is squelched, and the discourse of learning is closed. When a state of disconnection occurs, modernist teachers neatly resolve all contradictions and leave no end untied." (Kincheloe et al., 1999, p. 131)

We need to question the very existence of categorical dichotomies.
We need to question the basic ontological assumption that
the structure of the world is independent of the knower;
that the known is disconnected from the knower.

In pondering the subject/object dichotomy, I think it helps to see why external authority seems so important in this research with elementary preservice teachers. The modernist or Cartesian episteme is 'essentially' often defined and critiqued through the subject/object dichotomy. Postmodernist(s) (and) feminists 'centre' their critique of this dichotomy on the identification of its inherent hierarchies and phallocentrisms.

"In participating in the world of ideas as objects rather than as subjects we have come to take for granted that our thinking is to be authorized by an external source of authority." (Smith, 1991, p. 250)

> "Derrida reveals more clearly than any other postmodern figure that at the root of logocentrism is a set of inflexible binary oppositions. Informing all of these binary oppositions is the masculine/feminine opposition. Each of these oppositions is constituted by a hierarchy in which the privileged is associated with the masculine, the disprivileged with the feminine. Thus, as Derrida himself argues, logocentrism is phallocentrism." (Hekman, 1990, p. 26)

"Although both feminism and postmodernism are complex movements, the focus of both positions, at least in epistemological terms, is challenging the hierarchical dualisms of Enlightenment thought" (Hekman, 1990, p. 8). In terms of this subject/object dichotomy, I would like to propose that there are at least *three issues* in need of deconstruction and reconceptualization, or perhaps outright rejection. First, the dichotomy presents an autonomous, abstract subject (the knower) as distinct from a separate object (the known). Second, the subject's 'acquisition' of knowledge is grounded in the belief that there is an absolute and universal Truth. Third, the dichotomy (and indeed all dualisms based on Enlightenment thought) identifies the first element as masculine and the second element as feminine. Not only is this entrenched in essentialism but the privileging of the male (or first) element over the female (or second) element is a built-in feature.

Problem Solving

CRITIQUE...

that knowledge is acquired through the opposition of a knowing subject and a known object...
and imagine knowledge 'production' as a discursive process

that there is only one Truth...
and propose multiple truths

the privileging of subject over object...
and envision a reconceptualized relationship between knower and known

the gendered dualism of men as subjects and women as objects...
and challenge outright the layered hegemony of women's exclusion

If it is assumed that the 'features' of the Enlightenment subject/object dualism have been ingrained in Western society ***then*** it is necessary to face its epistemological implications: That men alone are the subjects who can constitute knowledge. Women are excluded from the possibility of be/coming the subject, the knower. "Because women are defined as incapable of producing knowledge, they are therefore defined as incapable of engaging in intellectual, and, specifically, scientific activities." (Hekman, 1990, p. 94) Feminist researchers approach this syllogism in multiple ways. One feminist approach is to leave the subject/object dichotomy intact and attempt to include women as subjects also (which leaves the Cartesian definition of subject more-or-less intact also). Feminist theorists from this perspective are more interested in gaining the opportunity to be/come subjects than to dissolve the notion of subject altogether (which is a focus of other feminist theorists). This approach could be thought of as leading to a *reformed* epistemology, rather than a *reconceptualized* epistemology.

> Feminism, like other forms of identity politics, is no mere celebration of difference. It is a response to perceived injustice. Power relations and power structures constrain who may belong in any social sphere. (Griffiths, 2006, p. 395)

My belief is that, unless the Cartesian definition of subject is problematized, then the space between subject and object, between knower and known, remains vacuuous and non-interactive. "Instead of recreating a patriarchal discourse based on establishing hierarchies, feminist and other ways of knowing embrace knowledge as a conduit that connects the knower and the known" (Kincheloe et al., 1999, p. 131). In other words, more than just the inclusion of women as knowers in the definition of subject is demanded—there must be a complete deconstruction of the dualism and its elements. To leave it intact is comparable, in my mind, to the familiar strategy of increasing the number of women in mathematics and science as a means to justify claims that the cultures of mathematics and science are not patriarchal or phallocentric. I do not think it is acceptable to continue teaching the principle that adding more women to the solution will neutralize it. The situation is far more complex than an acid-base neutralization experiment.

Two researchers in particular, Luce Irigaray and Julia Kristeva, are vehemently opposed to approaches that merely attempt to reform the subject definition so that women are included. Instead, they are concerned with disrupting, deconstructing, and reconceptualizing. [See **Inside Research** on the next page]

Recent poststructural and critical theorists studying identity issues of be/coming teachers write less about subject/object dualisms and more about how we come *to be* subjects: "A concern with subjectivity, or how we come to be subjects, can move us beyond essentialist notions of human nature, the ideal of possessing a noncontradictory self, and the push to embody these kinds of instabilities in spite of the complicated realities we live" (Britzman, 2003, p. 71).

Inside Research

with Luce Irigaray and Julia Kristeva

In simplistic terms, both Irigaray and Kristeva challenge the hegemony of, and reject, the Cartesian subject. Irigaray prefers to define woman as a being that defies dichotomies (and therefore epistemologies that produce them) because all dichotomies are both hierarchical and gendered. In rejecting the Enlightenment dualism(s), Irigaray states:

> For Being's domination requires that whatever has been defined— *within the domain of sameness*— as "more" (true, right, clear, reasonable, intelligible, paternal, masculine...) should progressively win out over *its* "other", its "different"-- its differing-- and when it comes right down to it, over its negative, its "less" (fantastic, harmful, obscure, "mad", sensible, maternal, feminine...). (Irigaray, as quoted in Hekman, 1990, p. 83)

Irigaray makes it quite clear that she moves beyond an analysis of woman as 'other' (as man's 'other'). She not only rejects the subject/object dichotomy in its hierarchical and gendered form but she rejects any attempt to see woman as subject *or* object, asserting that "any epistemology that is rooted in the subject is inherently phallocratic" (Hekman, 1990, p. 82).

In Kristeva's discussion of the subject, she asserts that "subjects are products of discourse, they do not exist in a pre-given sense, and they are not producers, but produced" (Hekman, 1990, p. 86). Unlike Irigaray, Kristeva is not rejecting the concept of the subject, but she rejects and replaces the Cartesian subject. Whereas Enlightenment epistemology assumes a *constituting* Cartesian subject, epistemologies explored by Kristeva propose a *constituted subject* (Hekman, 1990, p. 79).

> The reconstruction of the subject proposed by Kristeva constitutes a radical departure from the Cartesian subject because she challenges its constituting role. She replaces the constituting Cartesian subject with a subject that is constituted by discourse, but one that is by no means passive. Opposed to the fixed entity that is Descartes' knowing subject, she presents a subject in process, one that is constituted differently by different forms of discourse... she has so radically attacked the basis of the Cartesian subject that it is difficult to label the entity that she discusses a "subject" at all. (Hekman, 1990, p.91)

So, it seems, one theorist refuses to partake in any discussion of "subject" at all. The other theorist refers to "subject" but it is so radically different in meaning (where meaning is not fixed but in process) that it is hardly recognizable.

How do *you* view 'subject'?

Starting from the position that the Cartesian definitions of dichotomies relegate women to positions of epistemological inferiority, then we need to examine how "the displacing of the present patriarchal order of language will have both theoretical and practical political results" (Hekman, 1990, p. 92).

Returning to the research text at hand, I consider how this issue is more (or less?) problematic in my own research? I know that I cannot ignore the subject/object (or, indeed, any other) dichotomy because it will not simply be (re/dis)solved or changed by doing so. However, I still ask myself the question: if we continue to analyze and critique such dichotomies, do we risk merely leading to further reification of their current status? If we want to reject and/or reconstitute the subject(s) of knowing then let's get on with it.

Concept Review

- expose the imposed privileging of subject
- expose the paradigm of male as subject
- reject the privileging of subject
- decentre and reconstitute subject
- move on

At this point, I would like to introduce another angle of approach. What would happen if we placed less emphasis on the dichotomous elements themselves and more on the interaction between the two? In other words, place the privileging of the elements under erasure and, instead, reside in the messiness and ambiguity of the in-between space. The privileging of the dichotomous elements (good at / not good at; right / wrong; subject / object) ignores the space between—the 'slash' space that represents motion.

To tackle the long list of unsettling dichotomies that are present (often in their absence) in both the theoretical language of research and in the practical language of preservice teachers' perceptions of knowing, I will explore and implore the work of Ted Aoki on the concepts of metonymy and metonymical spaces of doubling.

In his insightful work on curriculum-as-plan and curriculum-as-live(d), Ted Aoki (1999) uses the word "plannable" to illustrate the concept of *metonymic spaces of doubling*.

> The word "plannable" invokes that which "is not", creating a split context as in plannable/unplannable. Contexturing this way brings forth the space between, here graphed with a slash: "/". It is in this space of between that our teachers, sensitive to both curriculum-as-planned and curriculum-as-lived, dwell, likely finding it a tensioned space of ambiguity, ambivalence, and uncertainty but simultaneously a vibrant site. It looks like a simple oppositional binary space, but it is not. It is a space of doubling, where we slip into the language of "both this and that, but neither this nor that". (p. 181)

METONYMY: IS / IS NOT

Reflexions

So, how does metonymy be/come part of my research? I will be re/reading Lacan's use of metonymy for a very long time and still wonder "what is he saying?" The figuratives of metaphor and metonymy both use concepts and words to "stand for another". In this text that I am writing, I use the metaphor of light because some of the properties of light resemble some of my participants' perceptions of knowledge and acts of knowing. Metaphor, from the Greek word *transfer*, is mainly used to refer to what is similar, drawing attention to likeness. Metonymy, on the other hand, demands attention to both likeness and unlikeness. Metonymy, from the Greek word for *change of name*, invokes both the word and the not-word. For instance, to speak about being "good at" inevitably calls to mind a state of being "not good at". Doesn't it? My feeling is that we are so well trained in the language of binary opposites that it is a s-t-r-e-t-c-h to imagine residing in the space between the poles. Am I right... or am I wrong?

"Doubling" is a term used to call to mind a form of deconstruction, where a hybrid is created that breaks away from the binary 'or' and moves toward a doubling 'and'. Metonymic space(s) of doubling create an image of two things happening simultaneously, such that there is a to-and-fro movement between awareness of the space (of absence) and a re-positioning to that new space. In the re-positioning, doubling is both a conjoining and disjoining.[4] Perhaps it is not so problematic to imagine, with this curriculum example of (un/)plannable, how one might dwell in a space that is "both this and that, but neither this nor that". After all, teachers experience and acknowledge (though perhaps uncomfortably) this ambiguous curriculum space on a regular basis—teachers are familiar with the to-and-fro movement between the lesson *plan* and the 'teachable moments' that, by definition, are *unplanned.* The metonymical space(s) of doubling are not so easy to envision, however, for other tensioned binaries, such as good-at/not-good-at or subject/object or presence/absence (or luminous/non-luminous or light/dark). Surely (she ponders aloud), these are either/or conTEXTS? Western epistemologies privilege ***good-at subject presence luminous light*** over ***not-good-at object absence non-luminous dark***. The use of metonymy, in the Aokian sense, is a desire to move away from privileging one over the 'other', and, in fact, even referring to an 'other'. The metonymical spaces are the spaces of ambiguity that represent denial of one real way, one truth. Aoki (2003) asserts that we have work to do in order to disrupt notions of truth and essence:

> …disciplined-oriented discourses of curriculum plans are grounded in the metaphysics of presence— privileging presence over absence. So valenced, the discourse assumes the presence of reality or truth hidden in the depth below, calling researchers to search and research, successful engagement resulting in findings that provide insights into the essence of reality. To research, then, is to represent the presence of the essence of reality. This is the language of the discourse of representation which in Western modernity has held hegemonic sway. (p. 3)

Deeply inscribed in the binarisms of Western epistemologies is the privileging of presence over absence, of subject over object, of light over dark. "One of the deepest convictions throughout Western philosophic and scientific thought is the belief that genuine knowledge or *theoria* is the most efficacious means for enlightenment and liberation of both the individual and society." (Bernstein, 1976, p. 176) The paradox, however, is "that what is frequently offered as genuine knowledge of social and political reality turns out to be a subtle form of ideology" (Bernstein, 1976, p. 176).

We (want to) believe in the presence, or pre-existence, of things but to believe in the 'already present' results in searching and re/searching for 'the real', for 'the truth'. But what can be found in the absence, in the space between? It seems to me that the space between is a space of possibility for growth and transformation. According to Aoki (1999), it is a "textured site of lived tension" but/and it is also "a generative site of possibilities and hope" (p. 181).

To deconstruct the binaries and their essences, we must first understand that deconstruction itself highlights both presence and absence, with long moments of uncomfortable ambiguity where we cannot always answer the question: "If not this, then what?" Perhaps a non-representational discourse is (present) in the absence. In being drawn to order, it is usually desirable to avoid messiness or slipperiness or confusion. How do we work and play in the metonymical spaces of doubling? How do we work and play (with enough LIGHT) in the metonymical spaces of doubling?

there is a crack,
there is a crack in everything
that's how the light gets in
(Cohen, 1998, p. 66)

that's how the light gets in

If/when we create these openings— these cracks for light to enter— how do we respond? Do we have a responsibility to respond? Do we try to shut it down, cover over the cracks, and suppress the light? Becoming aware of this (non-representational) discursive…

> **dis·cur·sive**
>
> **dis·cur·sive** (dĭ-skûr′sĭv) *adjective*
>
> 1. Covering a wide field of subjects; rambling.
> 2. Proceeding to a conclusion through reason rather than intuition.
>
> [Medieval Latin *discursīvus*, from Latin *discursus*, running about. See discourse.]
> (AHDEL, 1992)

space and how to act in this space will likely lead to discomfort. This discomfort, however, may open up new possibilities for transformation.

> But what if reality and truth can only be known by non-discursive thinking, a logic that twists and turns in many contradictory directions, a logic fuelled by the heart and imagination and dreams and hunches and feelings and indigestion, a writing and speaking and representing and reading and reflecting and researching that runs to and fro? (Leggo, 1999, p. 130)

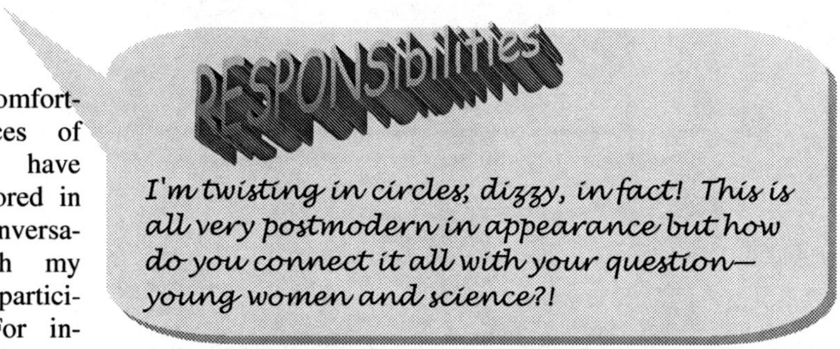

Such uncomfortable spaces of ambiguity have been explored in several conversations with my research participants. For instance, during a conversation with Ursula I asked her if she had ever experienced, or could even imagine, learning math *without emphasizing answers*. As will be seen in the next chapter of this text, Ursula was a very frustrated mathematics student because, for her, everything was "a race to an answer". To reside in a space of ambiguity or exploration would mean being left behind while the teacher and the other students raced to an answer. Even though such mad-dash math classes commonly frustrated Ursula, she asked me the question: "But how can you teach math without worrying about the answers?" To reside in a space of exploration and ambiguity seems unthinkable, especially since the culture of school is entrenched in (re)producing 'teacher-as-expert'— the one who knows the answer, with certainty. Gordon (2006) ponders the benefits of less certainty and more confusion in education, writing:

> … whereas the state of confusion may be psychologically uncomfortable or unsettling, educationally speaking it is not necessarily a bad thing. Indeed, confusion and uncertainty often lead people to search more, to questions previously cherished beliefs, and to take on a new perspective. (p. 16)

As educators, do we believe that the messiness and problematic nature of mathematics can actually be covered up if every question presented has its corresponding one right answer? To-and-fro movement between knowing and not knowing is what Ursula referred to as "getting lost in the middle," and the connotations are not good. My personal feeling is that "getting lost in the middle" might be exactly what Leggo (1999) means when he imag(in)es a knowing "fuelled by the heart and imagination and dreams and hunches and feelings" (p. 115). A knowing fuelled by the heart and imagination would mean, I imagine, learning to one's *heart's content* instead of being driven by *curriculum content.*

Metonymic Space(s)
of Doubling

LUMINOUS	**NON-LUMINOUS**
SUBJECT	**OBJECT**
PRESENCE	**ABSENCE**
GOOD AT	**NOT GOOD AT**
KNOWlous	**NON-KNOWlous**
LIGHT	**DARK**

Do you feel the
moment(s) when you
pass through the
slash? One usu-
ally hopes the
process is not
o s m o t i c —
slowly, porously mov-
ing from non-
privileged absence
to privileged pres-
ence, residing in the
messy, ambiguous
inter-spaces for what
seems like forever.
No, it is much more
desirable and accept-
able to be ***either/or.***
This slash space is
uncertain for me, and
I find I cannot label
things clearly— the
'here' OR 'there' *be/*
comes ***(t)here.*** *But*
still I wonder: in
theorizing these bi-
nary oppositions,
did anyone ever
question whether
they (il)luminate dif-
ferences al-
ready present
(not absent)
or are they respon-
sible for creating
them?

Where Does Light Come From?
Vocabulary Review[5]

Review CLUES...

Across

2. part of the house for general leisure activities (**not** learning)
4. self-knowledge
6. 'change of name' invoking is/is not
8. not eastern
9. having the truth and knowledge
10. wandering or shifting from one subject to another
11. conjoining and disjoining
12. the fact or condition of being at hand; non-absence

Down

1. a division into two parts (is; is not)
2. self-light
3. dealing with facts and objects, not thoughts and feelings
5. intentness on objects external to mind (and body)
7. requiring or showing special knowledge

How does light travel?

Part of teaching is helping students learn
how to tolerate ambiguity,
consider possibilities,
and ask questions that are unanswerable.
(Lightfoot, 1999)

CHAPTER TWO

HOW DOES LIGHT TRAVEL?

Part I: Rectilinear Propagation of Light

"Light is always traveling—there is no static light resting somewhere." (Duit, 1995, p. 276)

The rectilinear propagation of light is the technical terminology applied to the principle that 'light travels in straight lines'. (Jenkins & White, 1976, p. 4)

preVIEW...

This chapter begins with a critique of the traditional transmission model of knowledge with/in education. Arguments are presented for the necessity of (inter)acting with knowledge, thus moving away from such a transmission model. Theories of constructivism and conceptual change models are presented to provide context for a few of the ways in which traditional transmission models of teaching and learning are being challenged. Despite the overwhelming number of studies advocating the practical implementation of these theories in mathematics and science education, little seems to have changed. Why? To address this question, I explore the feasibility of expecting such changes to take place with/in a cycle of (mis)education. Beginning with preservice teachers' experiences of mathematics and science, I discuss a few of the barriers inherent in attempts to interact with knowledge in different ways. I propose a deconstruction of the traditional ways of see(k)ing 'answers' and propose a reconstruction of other spaces for realizing a pedagogy of purposes and interactions. Teaching and learning in these new spaces requires a rejection of the image of knowledge travelling straight line paths between a luminous source and the illuminated other— between **the** question and **its** answer. The spaces are frequently ambiguous and volatile, and call for a (re)positioning that is often uncomfortable and messy for both teachers and students alike.

Does knowledge travel in straight lines, from teachers' minds and textbook pages to students' minds? Advocates of constructivism have long recognized that *construction*, not *transmission*, of knowledge is necessary for learning. Almost seventy years ago, Dewey (1938) criticized the underlying principles of traditional education when he stated: "The subject matter of education consists of bodies of information and of skills that have been worked out in the past; therefore, the chief business of the school is to transmit them to the new generation" (p. 17).

The experiences of my preservice teacher research participants in this study strongly suggest that efforts to 'transmit' knowledge are not obsolete practices in the teaching of mathematics and science. While it may be desirable to believe that theories supporting constructivist learning have made ideas of knowledge transmission obsolete, it is apparent that few of the learners with whom I conversed actually experienced these constructivist theories in action.

A Matter of Perception...

"Left alone, light travels in straight lines." (Hecht, 1987, p. 5)

The phenomenon of light travelling in straight lines calls to mind, for me, an unfortunate, misinformed lecturer who thinks:

If I stand at the front of the classroom, as the source of knowledge, the continually emitted bits and pieces of my knowledge will travel in straight lines in all directions until they eventually collide with, and are absorbed by, my students, who sit passively awaiting illumination from within their darkened state of unknowing.

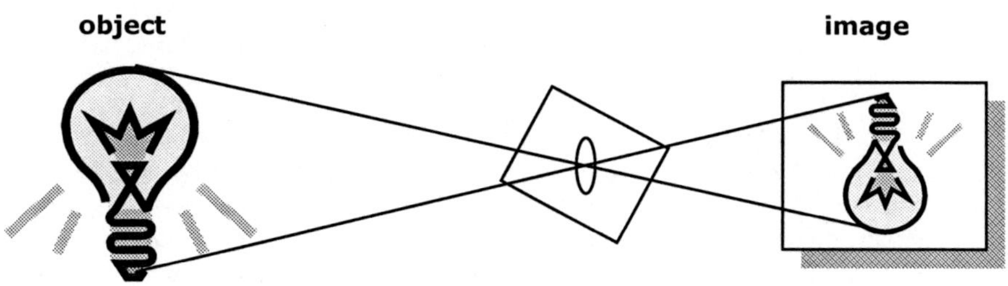

everything
everything
we followed our textbook
100%
especially in elementary
I saved all my notebooks and got a copy of the textbook and we looked...

it was just coming
RIGHT OUT OF THE TEXTBOOK
see the notes
copy the questions
find the answers

(Preservice teacher; pilot study April 1999)

We labelled diagrams that were photocopied out of the textbook.
I think we coloured them.

At least we had a science room.
We would mostly just sit there and not draw on the desks.
It also doubled as the lunchroom.

(Evelyn, December 1999)

> Our main premise is that significant learning is likely to occur only if the "facts" to be learned are construed by the learner as having personal relevance. We suggest that a "cultural transmission" approach to teaching and knowledge dominates science education. This approach has neglected the role of students' personal experiences in their construction of knowledge. (Pope & Gilbert, 1983, p. 193)

Bruner (1996) echoes my concern that there is support for the rectilinear propagation of knowledge as learning. He states: "What is to be learned by the pupil is conceived as 'in' the minds of teachers as well as in books, maps, art, computer databases, or whatever. Knowledge is simply to be 'looked up' or 'listened to'." (p. 55)

What, then, does it mean to learn? Roth & Barton (2004), in discussing alternatives to traditional science education, propose that "learning is the expansion of possibilities for acting in and toward the world" (p. 9). Ernst von Glasersfeld (1995) asserts that in order for learning to 'happen' (note the *action* verb), students must acquire a sense of ownership in what is being taught and learned; this, he claims, is one of the most important tenets of constructivism

> From a constructivist perspective, learning… requires self-regulation and the building of conceptual structures through reflection and abstraction. Problems are not solved by the retrieval of rote-learned "right answers". To solve a problem intelligently, one must first see it as one's own problem. That is, one must see it as an obstacle that obstructs one's progress toward a goal. (p. 14)

The traditional models of teaching, where the students are viewed as passive learners and receivers of knowledge, cannot pretend to instil this sense of ownership in the students. Students must be encouraged to construct and interact with knowledge in attempting to make sense out of their experiential worlds.

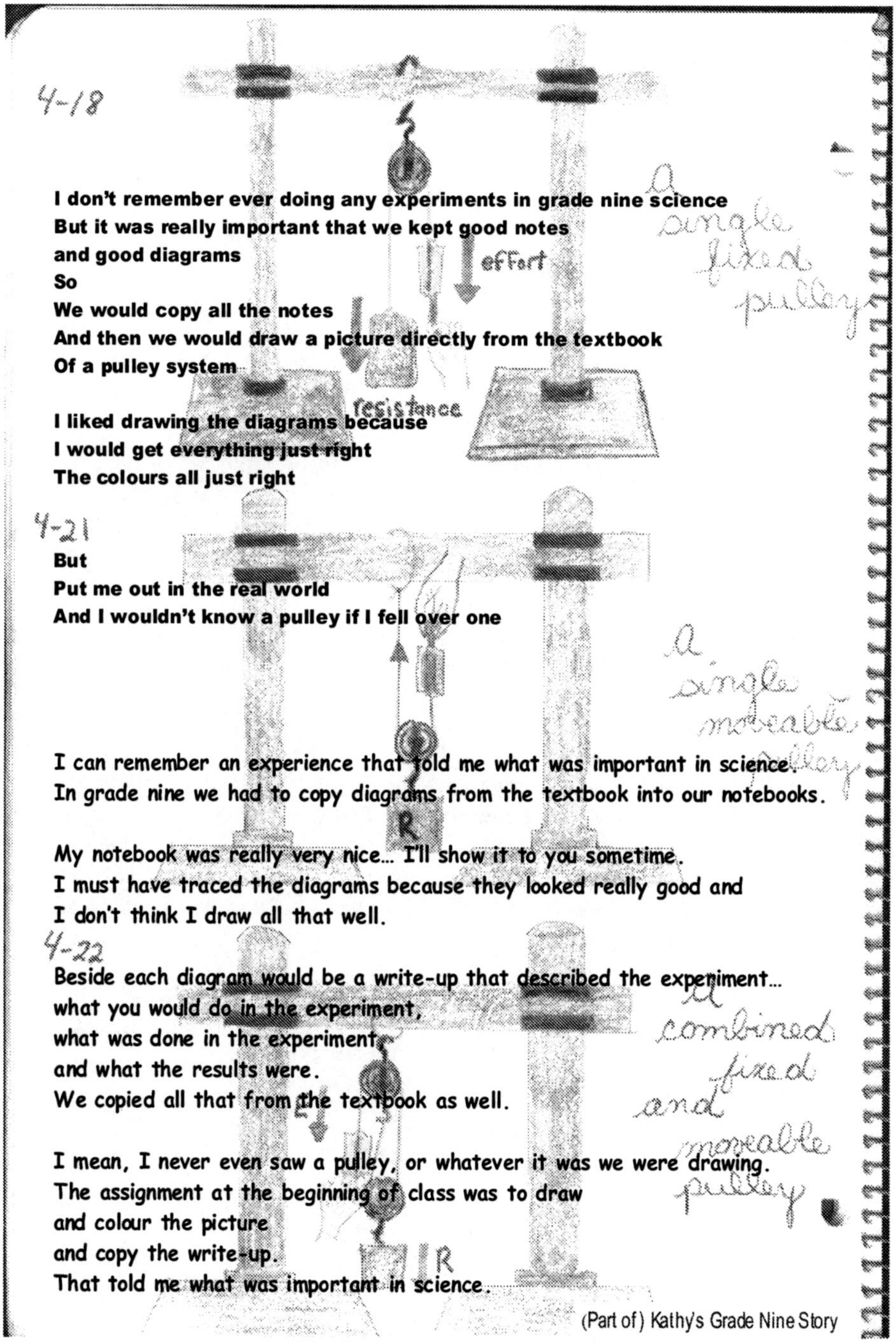

I don't remember ever doing any experiments in grade nine science
But it was really important that we kept good notes
and good diagrams
So
We would copy all the notes
And then we would draw a picture directly from the textbook
Of a pulley system

I liked drawing the diagrams because
I would get everything just right
The colours all just right

But
Put me out in the real world
And I wouldn't know a pulley if I fell over one

I can remember an experience that told me what was important in science.
In grade nine we had to copy diagrams from the textbook into our notebooks.

My notebook was really very nice... I'll show it to you sometime.
I must have traced the diagrams because they looked really good and
I don't think I draw all that well.

Beside each diagram would be a write-up that described the experiment...
what you would do in the experiment,
what was done in the experiment,
and what the results were.
We copied all that from the textbook as well.

I mean, I never even saw a pulley, or whatever it was we were drawing.
The assignment at the beginning of class was to draw
and colour the picture
and copy the write-up.
That told me what was important in science.

(Part of) Kathy's Grade Nine Story

Quote of the day...

They gave you a handout;
you memorized it quick the night before
and you go spit that out and forget it the
next day.
(Ursula, January 2000)

"Left alone, light travels in straight lines."
(Hecht, 1987, p. 5)

Noddings and Shore (1984) write about engaging the Will: "... any arrangement or presentation of subject matter... must somehow function to secure the active participation of the student. The student must be involved in the *construction of purposes*" (p. 123). Noddings and Shore suggest that students are often split into an obedient part and a rebellious part when presented with subject material to be learned. In the following quote, they speak about healing this split:

> Clearly, the student cannot receive the material fully while his or her attention is directed at himself or herself thinking: Why am I doing this? Why should I do this? How can I get out of this? The very first task of both curriculum maker and teacher, then, is to engage the Will, to win that initial commitment. (Noddings & Shore, 1984, p. 123)

To win this commitment, however, I believe there is a need for a QUEST in these QUESTions of purpose. "There is, I think, no point in the philosophy of progressive education which is sounder than its emphasis upon the importance of the participation of the learner in the formation of the purposes which direct his activities in the learning process..." (Dewey, 1938, p. 67). Dewey further states that the formation of a purpose always begins with an impulse and a desire. What more appropriate way might we demonstrate the spark of our impulses and desires than by asking questions: How? When? Why? With my participants, this QUEST for purpose and ownership in their learning was often expressed in their desire to know 'why.' Too frequently, however, their QUEST was also quashed in the teacher's desire **not** to be asked 'why'.

It really stuck with me that I always wanted to know why.
I was really compulsive about it. It bothered people. I had trouble with algebra
when I first started. So, I'd ask my friends.
No one could tutor me because they'd say
'"Well, now you cross out the '1' and carry the 'x'...", or whatever,
and I'd be like "why?".
"You just do."
"Well, why?"
They all gave up and passed me on to someone else. It was really hard.
My teachers would say
'Go away, you just do'.
Nobody understood that I had to know why or else I couldn't do it.
I thought,
'I'm stupid. Obviously this is just what you have to do.'
I don't know.

I don't know if other people just don't *think* about why you're supposed to be doing this or if they have an *understanding* that I don't have.
(Helen, April 2000)

"Engaging the Will" and *construction of purposes* stress the necessity for learners to interact (in a meaningful way) with the subject material. Knowledge does not become one's own if this interaction or participation (in the form of purpose, choice, commitment, engagement) is missing. How can the rectilinear propagation of knowledge ever bring about learning? How is it that we can expect students to read a problem from a textbook and be committed to searching for its 'answer'?

> The effective motivation to continue learning can be fostered only by leading students to experience the pleasure that is inherent in solving a problem seen and chosen as one's own. (von Glasersfeld, 1995, p. 14)

> At every level of possible understanding, it is essential to offer, even to demand, choice. By choosing, selecting, employing, students engage the intuitive functions that will work in their quest for meaning. (Noddings & Shore, 1984, p. 149)

Seeking Desire and Pleasure...

It is interesting to note how educational researchers like John Dewey, Jerome Bruner, and Ernst von Glasersfeld use words like *pleasure* and *desire* when writing about learning. Carl Leggo and Ted Aoki, in writing about the important distinctions between dichotomy and metonymy, suggest that we are always moving to-and-fro in the dance between thing and no-thing— between pleasure and frustration, for instance. It is not about choosing pleasure OR frustration but in residing in the ambiguous space(s) *between.* In learning, we need to expect, and even welcome, this continuous movement between pleasure and frustration, not clearly one or the other, but both. A sort of ***pleasurefrustration*** way of being in this world. Perhaps we (as educators and learners) spend too much time avoiding the frustrating aspects of learning— thinking that by handing over the 'new thing' on a platter, the learner will experience less frustration and displeasure in the learning process, thereby leading to a better learning experience.

In deciding whether something is worth knowing, Bruner (1979) suggests applying two main criteria: "Whether the knowledge gives a sense of delight and whether it bestows the gift of intellectual travel beyond the information given…" (p. 109). In pondering the question of what produces pleasure in a person's intellectual life, Bruner responds that "pleasure is found in the act of reducing surprise and complexity to predictability and simplicity" (Bruner, 1979, p. 110). To me, this speaks of reaching a point of understanding and meaning in learning which gives one a sense of pleasure and delight in having set out on the journey. The importance of this thread of delight, weaving in and through all acts of learning, is often overlooked. How does one instill a sense of ownership, purpose, desire and delight in acts of learning?

Reflexions...

Growing up, my friends and I would quiz each other on definitions and essays. We would proudly boast of *knowing our schoolwork by heart*. "Go ahead. Ask me. I know it off by heart." I certainly experienced purpose, desire, and delight in this act... but was it *an act of learning*? Well now, that would be the heart of the matter, right?

Teachers cannot pretend to know all of the "intellectual travels" of their students, in the form of both past journeys and future desired expeditions. It is possible, however, to recognize the insight into learning provided by Ausubel when he states that "the most important single factor influencing learning is what the learner already knows" (Ausubel, 1968, p. iv). Attempts at knowledge construction will fail if one does not first take into consideration the foundation of experiential knowledge already in place. The constructivist view is that students must make sense out of something new in conjunction with their existing knowledge. A common feature of the literature on constructivism advocates that teachers be concerned with what goes on in the student's head, how they are making sense of their world and why they believe their ideas will lead to a solution instead of focussing on the solution itself. And, as Skovsmose (2005) proposes regarding the learning of mathematics:

> Students become recognized as the constructors, and the only possible constructors, of their mathematics knowledge. Like any knowledge, it must be homemade. As a consequence, many suggestions have been made to ensure a

> learning environment to facilitate the students' construction of their own knowledge. The development of notions such as the 'constructivist teacher' and 'constructivist mathematics education', make sense. They refer to a teaching practice that acknowledges that students do construct their own knowledge on the basis of what they already know. (pp. 153-154)

The behaviourists' approach of fine-tuning examples, algorithms, drills and reinforcement to achieve desired results without emphasizing understanding has received much criticism from the cognitively oriented community of researchers (von Glasersfeld & Steffe, 1991), leading to some very interesting research on alternative conceptions and conceptual change models in teaching.

Inside Research...

The literature on alternative conceptions reinforces the notion that students come to the classroom with prior knowledge which influences further learning and, at the same time, is resistant to change (Wandersee, Mintzes & Novak, 1994). These conceptions are firmly rooted in their life experiences and as long as they seem viable, students see no reason to alter or change them in any way. Posner, Strike, Hewson and Gertzog (1982) propose a model for conceptual change in which one of the conditions for change to occur is that students must first experience dissatisfaction with their models before they will consider an alternative one. From the teacher's perspective, one cannot help a student arrive at such dissatisfaction unless the student has expressed his/her thinking or way of knowing first. A teacher must have some idea of the conceptual structures with which the students operate in order to intervene in the process of conceptual development or modification (von Glasersfeld & Steffe, 1991). Only then can instruction be guided toward the introduction of anomalies that challenge the viability of the student's model. Several researchers have written on the subject of alternative conceptions and conceptual change theories (e.g. Duit & Treagust, 2003; Hewson & Thorley, 1989; Schoon, 1995; Tillema & Knol, 1997; Driver & Erickson, 1983), with the common thread in the literature being that the students' models must be verbalized and then challenged before there is any chance that a more plausible model will be accommodated.

Much of the research on conceptual change is based on rational thought processes and not strategies for facilitation (Beeth, 1998). As Wandersee et al. (1994) point out, there is no shortage of alternative conceptions research that examines elementary school teachers and elementary school preservice teachers. This is understandable, under the circumstances, since science educators are concerned about the adequate preparation and abilities of elementary teachers. After all, these are the teachers who will be required to develop and implement conceptual change instruction, as called for in the literature. Two questions emerge as possible obstacles. Firstly, are we asking teachers to address the alternative conceptions of students without first asking if anyone has addressed *their* alternative conceptions?

To expect a teacher to use methodologies in the classroom that take into account alternative conceptions without the teacher first confronting her/his own conceptual understanding (and potential alternative frameworks) is like putting the cart before the horse. Students in teacher education programs are likely to construct models of teaching and knowing that fit with their own conceptual frameworks, constructed during many years of learning in traditional classrooms (Stofflett & Stefanon, 1996). Several teaching scenarios are possible, with the worst case being that the teacher only reinforces (unknowingly) her/his *own* alternative conceptions in the students.

Secondly, assuming the teacher recognizes the need to address students' alternative conceptions, the question is whether s(he) will have the content knowledge (both in depth and breadth) necessary to do this? Smith and Neale (1989) ask what kinds of knowledge are required for successful conceptual change teaching. These authors focus on two categories of teachers' content knowledge: substantive and pedagogical.

The first category (substantive knowledge) requires, in addition to just factual knowledge, the knowledge of conceptual relationships and applications. The second category (pedagogical knowledge) receives much more attention in teacher education programs, since it involves the knowledge of particular teaching strategies that will assist students with conceptual understanding. Though Smith and Neale carefully describe teaching strategies related to the second category of knowledge, they fail to understand why elementary teachers so rarely use these strategies, even in light of the importance they place on substantive knowledge. The authors do not seem to make the connection between a preservice teacher's substantive knowledge (or lack thereof) and her/his desire to risk residing in foreign and uncomfortable pedagogical spaces in the classroom.

sub·stan·tive (sùb'sten-tîv) *adjective*
Abbr. **s., sb., subst.**

1. Substantial; considerable.
2. Independent in existence or function; not subordinate.
3. Not imaginary; actual; real.
4. Of, or relating to, the essence or substance; essential: *substantive information.*
5. Having a solid basis; firm.

ped·a·gog·ic (pèd'e-gòj'îk, -go'jîk) also **ped·a·gog·i·cal** (-gòj'î-kel, -go'jî-) *adjective*

1. Of, relating to, or characteristic of pedagogy.
2. Characterized by pedantic formality: *a haughty, pedagogic manner.*

(AHDEL, 1992)

Stofflett and Stoddart (1994) state things quite simply when they stress that in order to *teach* conceptually, the teacher must *understand* the content conceptually. Studies have shown (Hashweh, 1996) that teachers can discover student alternative conceptions only in the areas of science in which they themselves were experts. It is a tremendous risk-taking endeavour for a teacher with weak substantive content knowledge to explore student ideas, to engage in scientific inquiry, and to encourage debate and discussion over discrepant events. In fact, I would maintain that a teacher who is a product of the didactic learning-teaching cycle has many skills to acquire before breaking out of this cycle— skills such as directing students to think about and explain their conceptual structures (developing metacognitive skills), determining the level of commitment a student has to their alternative conceptions, and introducing anomalies or counterexamples to help refute these alternative conceptions (Beeth, 1998; Posner et al., 1982). These skills are not easily shown or explained but, as with most skills, need much guidance and practice.

> It is tempting to suggest that this 'cyclic' state of affairs could be remedied by placing the teacher in an environment in which new content could be learned in a manner facilitating understanding. The implication is that teachers' beliefs are causing the curriculum to be the way it is. (Tobin, Tippins & Gallard, 1994, p. 52)

The above quote suggests a dialectic relationship between teachers' beliefs and curriculum implementation. Within the literature on preservice teachers and alternative conceptions, there are attempts to examine more closely the relationships between attitudes and beliefs of preservice teachers toward science and mathematics and their planned pedagogy for teaching these subjects. Studies have shown (Kurth, 1995; Riggs, 1991; Roychoudhury et al., 1995) that many students enter elementary teacher education programs with an obvious dislike and disinterest in science, both for themselves personally and in the prospect of teaching the subject.

> "In classical philosophy, **dialectic** (Greek: διαλεκτική) is an exchange of *propositions* (theses) and *counter-propositions* (antitheses) resulting in a *synthesis* of the opposing assertions, or at least a qualitative transformation in the direction of the dialogue." (http://en.wikipedia.org/wiki/Dialectic)
>
> **IN OTHER WOR(L)DS**, "... teaching and learning science needs to be seen in dialectical terms. As a tool, science can be used both for and against a particular position over a contentious issue. But more importantly, science is only a good tool if it can be used reflexively, that is, to critique and even deconstruct itself." (Roth & Barton, 2004, p. 15)

> Science teachers often move through 15 to 20 years of schooling without ever being induced to think about their own beliefs about the nature of science and scientific knowledge and that which has shaped them... understanding teacher beliefs can lead to redefinitions of our images of science and the way it is taught. (Tobin et al., 1994, p. 62)

Similarly, in mathematics methods courses, most preservice teachers view mathematics as a set of rules and procedures to be memorized (Hill, 1997). Burton (1996) states that "the social nature of mathematics remains a well-kept secret especially amongst the wider lay public who firmly believe that mathematics consists of a body of 'objects', or facts and skills, universally recognizable, and transmissable within a classroom setting (if only teachers were more effective)" (p. 31).

This public belief, in addition to students' experiences of didactic teaching, "has led to a socially differentiated distribution of those who learn it successfully, and those who do not" (Burton, 1996, p. 31).

The beliefs teachers hold about mathematics are the result of these skill- and rule- based experiences prior to teacher education programs and, without significant intervention, change seems unlikely. The question of the specific nature of this intervention is the subject of much of the research (Cooney, Shealy, & Arvold, 1998; Kang, 1996). The traditional view of mathematics as a well-defined and well-described body of knowledge, rooted in the belief that it has been discovered rather than invented, has led to the idea of mathematics knowledge as a commodity that can be transferred from teacher to learner (von Glasersfeld & Steffe, 1991). Of the need for developing an understanding of conceptual relationships in science and mathematics, von Glasersfeld writes:

> Science, having to a large extent replaced religion in the 20th century, is all too often presented as the way to absolute truth. If mathematics were explained as a way of operating with a particular kind of abstraction and science as a way of building models to help us manage the world we experience, some of the latent resistances might be allayed. But this, again, would require some delving into conceptual foundations." (von Glasersfeld, 1995, p. 5)

Based on conversations with my research participants, it is clear that teachers' beliefs and attitudes have more of an impact on science teaching and learning than we might imagine. In any attempt to examine the beliefs and attitudes of preservice teachers, however, a researcher cannot ignore a few crucial points. First of all, the majority of elementary teachers are female (Roychoudhury et. al., 1993). In examining female preservice teachers' beliefs about the nature of science and of constructing scientific knowledge, we cannot dismiss the androcentric influences of science that have excluded spaces for more alternative and less positivistic ways of knowing for so many years of schooling.

"Didactic teaching usually is based on the notion that pupils should be presented with facts, principles, and rules of action which are to be learned, remembered, and then applied." (Bruner, 1996, pp. 54-55)

Secondly, the studies direct one's attention to the fact that most teachers' attitudes and beliefs stem from a lack of self-confidence in their ability 'to do' science and mathematics. Many preservice teachers have past educational experiences immersed in a didactic approach, emphasizing mostly procedural knowledge. Roychoudhury et al. (1995), in proposing that ownership of learning, personal connections, and a feeling of self-efficacy are essential to female learners, indicate that the traditional didactic classroom is likely to have only negative overall outcomes. The impact of these prior experiences for female preservice teachers is that the negative outcomes are often translated into anxiety and a lack of self-confidence in science and mathematics and the teaching of these subjects.

If you start elementary school off with kids learning
how to be curious,
how to find answers out,
how to do explorations and design their own experiments
and if you could just carry that all the way through…

I don't know but I think probably secondary teachers
are trained differently from elementary teachers.
I know the courses they take are a lot more *content*-based.
We learn more about the *foundations* for teaching science
and **they** learn more *content.*

I think that's part of the problem.
Maybe they should come and sit in our classes and watch what we're doing.
(Helen, April 2000)

While many of the attitudes and beliefs of preservice teachers can be traced back to didactic pedagogy and conceptual knowledge fragility, conversations with my participants strongly suggest that the solution is not simply to address content and methodology issues as a means of effective teacher education. Female elementary preservice teachers may be the first to acknowledge that more mathematics and science content taught the same old way is unlikely to improve the situation (Stofflett & Stoddart, 1994).

See the enclosed Newsletter (pp. 86-87) entitled
Knowing Eclipses (and) Eclipsing Knowing
for a focus group discussion on knowing (in) science

> If women have to teach in an area that is incompatible with their ways of knowing and comes in a masculine packaging then, it is no small surprise that they will avoid the subject in their own learning as well as in teaching. (Roychoudhury et al., 1993, p. 37)

Let us look, for a moment, at a particular example in science. Imagine an elementary teacher introducing a student-centred inquiry into the causes of the seasons. The teacher feels conceptually prepared to facilitate the inquiry process—she understands the significance of the tilt of the earth's axis, the earth's rotation about the sun, and the resulting seasons experienced in various parts of the world. In other words, she understands the *why* of this scientific phenomenon. The question is, however, does she understand the *why not* of each of the student's alternative conceptions? For example, does she know why the tilt of the earth's axis does not change (even though the physical model of the globe does not have a fixed axis), or why the distance differential between the earth and the sun at certain times of the year is not responsible for the seasons (how many times have we heard the season of winter explained in terms of 'the earth being farther away from the sun' at this time of

year?)? This teacher must decide if the risks of constructivist teaching aimed at conceptual change are worth it. [After all, as a teacher, it **is** a risk to admit to not knowing, isn't it?] In order to effect conceptual change in students, the teacher must diagnose errors in student thinking, help create cognitive conflicts in students through a discrepant event or activity, and demonstrate the fruitfulness of the new model (Hewson & Thorley, 1989). Compare this with the traditional teaching model of 'information transfer' from teacher and textbook to student. If you are a product of the latter model of science pedagogy and are anxious about science content, how would you be inclined to teach?

"Left alone, light travels in straight lines."
(Hecht, 1987, p. 5)

To emphasize other ways of knowing and to value process over product, a teacher must be willing to take risks in the classroom—that is, to become comfortable with uncomfortable in terms of one's conceptual knowledge and in terms of student questions. Confusion, ambiguity, and uncertainty are not to be viewed as forbidden places to reside; in fact, I suggest that they are necessary for that *pleasurefrustration* way of being in this world. Attempting to convince us of this, Gordon (2006) states:

> … whereas the state of confusion may be psychologically uncomfortable or unsettling, educationally speaking it is not necessarily a bad thing. Indeed, confusion and uncertainty often lead people to search more, to questions previously cherished beliefs, and to take on a new perspective. (p. 16)

While it may feel uncomfortable and 'risky' for teachers to adopt non-traditional, less didactic pedagogy, it is also likely that such approaches carry more potential and possibility for learning. What are some of these possibilities, along with their associated risks? The possibilities that I am about to discuss require a re-positioning with/in pedagogical practices. As I see it, from my own experiences and from those expressed in the voices of my participants, there are three 'risky' pedagogical spaces that call for exploration. They are 'risky' spaces in that they are uncomfortable and messy, for both teachers and students alike. The spaces do not focus on issues of content knowledge (as so much of the recent research does) but on re-organizing and re-ordering already existing knowledge through a re-positioning of perspectives on *what it means to know*.

★ The space(s) of the 'question'… how might we ask questions differently and ask different questions?

★ The space(s) of the 'answer'… how might we begin to recognize the importance of encouraging different paths leading to different destinations?

★ The space(s) in/between 'question' and 'answer'… is it possible to (re)side in the space between, experiencing a to-and-fro movement between question and answer, which emphasizes a journey not a destination?

In the first space, the focus is on asking questions differently and asking different questions. A traditional didactic model of teaching often seems modelled after the game of *Jeopardy*. There is one right answer and only one possible question that could be asked to get you there— arriving at the question is really no more than working backwards from the answer. Several of my research participants commented exactly on this— that a common approach to mathematical problem solving is working from the answer section in the back of the textbook.

> You can figure out what they want you to do.
> You can do it your own way and then fill in the blank with the answer
> cause the *answers* are in the back of the book, right?
>
> I used to work from the answer all the time.
> Like, get the answer from the back of the book then,
> Once you have that answer,
> you almost have the confidence to do the question.
>
> (Ursula, January 2000)

> **RESPONSibilities**
>
> Unfortunately, our "system" requires (expects) some kind of evaluation to provide some account of the learning! How do you carry your thesis of spaces into evaluation? Why must we account for learning in such archaic ways?

The image seems to be that the shortest path between question and answer is a straight line, and it is expected that everyone will tow that line. Now, I am not suggesting that this form of pedagogy is so common because teachers are deliberately trying to be inflexible. As mentioned previously, *it may be the only way they know* based on years of traditional, didactic pedagogy. To ask questions differently and to ask different questions is to step out of the familiarity of our comfort zones; it is, however, an important, yet often overlooked, aspect of teaching differently. In fact, the other two spaces, which deal with process and product (re)construction, make little sense if we do not first consider how we might radically change our ways of questioning. I believe that different ways of knowing are closely connected to different ways of asking.

I do not ask the question in order to answer the question;
I ask the question, again and again,
in order to know the question.
My research is really about opening
my ears and eyes and tongue and skin and nose and lungs and
heart and spirit...
(Leggo, 1999, p. 113)

Knowing Eclipses (and) Eclipsing Knowing

Research Newsletter
January 27, 2000
Volume 1, Issue 1

Preservice Teachers Participate in Focus Group Discussion

A small group of preservice teachers got together to discuss their reactions and thoughts on the following scenario:

A person from your community, someone you don't know very well, discovers that you are a teacher. She is confused about something in science and says 'oh, I should ask you, *you'll* know the answer.' She is trying to figure out what the lunar eclipse is all about... what causes it and why it doesn't happen very often. So she comes to you with the question.

Roberta: I would be honoured to be approached... that someone thinks that I have this knowledge base that I could share. And I think we don't remember that as teachers... that people expect that we have a greater ground of knowledge.

Evelyn: I don't think that's necessarily true. I think as a teacher they expect you to know everything... I just think you forget to feel honoured about it. (laugh) My response to this would be to give the basic definition of a lunar eclipse, saying I don't really know much more than that, and start pulling out everything off the shelves... 'oh, well here's this *really* great book about lunar eclipses' and then I think I'd just *harass* this poor lady for weeks with information about lunar eclipses as I kept getting it and getting it. But the thing is, in terms of my scientific knowledge, it only comes into play when somebody *else* needs it. I don't really care...

Sylvia: It sort of sounds like you're saying— maybe I'm wrong but— 'ok, initially I don't really *know* and so here's a book.' And then afterwards you're proving that you *do* know. Like you have to study up hard on it but then for the next two weeks you'll be proving 'oh, yeah, I know... I know *exactly* what you're talking about.'

'Here's this *really* great book..'.

Harriet: All I would think is 'oh, crap.' I think that's part of... well, beforehand she maybe thought I was smart enough to know this [but] because I could only maybe give a *very* basic definition, if that, I would think she's walking away going 'yeah, *she's* qualified to teach...', thinking I was a moron pretty much. If I couldn't give her a definition, I'd be apologizing for it. 'I'm so sorry, I'm so sorry.' Like, I *love* what you [Roberta] said about being honoured, because that's the *last* thing that would ever come to my mind. I'd just be like 'oh dear, please ask someone else' because I'd be thinking 'what's her opinion now of me?'

Elsie: It'd be like she was identifying a weakness in me and I would want to go out and find all the information so that I didn't *have* that weakness anymore.

Special points of interest:

- ✓ Seven participants involved in discussion group
- ✓ Preservice teachers discuss how they feel about their knowledge of science
- ✓ Public image of science and science teachers is reflected upon

Volume 1, Issue 1

Maude: My initial reaction was like 'whoa'. We talked about how it would peak your interest and you'd want to find out why and maybe you could find out together. Together you could talk about what *you* know, talk about what *they* know, and then go get some more information.

Sylvia: That sounds great for *kids* but I wouldn't do that with an *adult.*

(sounds of agreement among participants)

Sylvia: With my kids, 'sure, yeah, let's find out more about that' but with an adult, it's like— especially if they're older than I am, they're *testing* me so— no, I couldn't do that with an adult. I'd give them a brief definition and then, like [Evelyn] said, give them a book, and then find out everything I could possibly find out and overload them with information until they're sick of it. Yeah, I couldn't do that with an adult— unless I knew them really well that I was personally comfortable with them. But if I didn't, no way.

Maude: Do the old: 'I'm not sure... can I get back to you tomorrow?' (laugh)

Elsie: 'I'm *really* busy right now.' (laugh)

Sylvia: (laugh) 'I have tons of homework to correct.'

Kathy: My reaction might be to pretend that I just couldn't deal with it then. Because I have actually had people respond, when I don't know the answer they're looking for, 'don't you have a masters degree?'

Sylvia: Oh no. If I *ever* did that I apologize! (laugh)

Kathy: Thank you. (laugh) As you get more and more specific in studying something, people tend to think that your knowledge becomes more and more global... I don't think it works that way.

Helen: Well, my reaction is exactly the same as [Sylvia's]. If that were my kids then I'd be like 'ok, lets' find out, that'll be fun.' But if it's another adult, someone older, I would just be 'oh my god, I don't want to answer you, go ask someone else.'

Kathy: Do you think that goes back to what the public thinks about *you* or does it go back to what you think about your own knowledge?

Helen: I think the public...

Sylvia: Yeah, I think they see the teacher as 'well, they should have some degree of knowledge, [they've] been here for four years'. We should know something.

Harriet: I think there's a lot of scrutiny with teachers, and especially now with so many... marks being published... you're *just scrutinized*. I take it fairly seriously... and not being able to answer questions... 'Well, I *told* you, that's why our kids aren't learning science, they're doing poor in math... it's because *you* can't do that.' It's always that, for me, it's back to 'you'. I think that's part of the problem.

Elsie: I think it's *both.* Of course, I don't want the public to have a poor opinion of me or other people to have a poor opinion of me, but I would like to think that I would know that. And I think that I should. But I don't.

"...I would like to think that I would know that."

Kathy: Is that because of the particular scenario that I chose or do you think you'd feel this way with other things also?

Sylvia, Elsie, Harriet: A *lot* of things.

Roberta: I think the reality is there's *so much* out there, *so much* knowledge, that our human brains are just not able to retain it.

The second space (the space of the 'answer') in a risk-taking pedagogy involves encouraging students to explore the (pro)posed questions in various ways and propose different models for 'answering' them. Of course, there are (risky) implications for the teacher because s(he) is faced with the task of determining the viability of the students' various models. It is often a challenge, in general, to understand another person's conceptual framework, let alone to identify alternative or mis-conceptions and then teach for conceptual change (as discussed previously in **Inside Research**, on p. 79). For teachers to identify and legitimize *other* viable processes (in addition to the one they may be most comfortable with) means accepting multiple ways of knowing and expanding our notions of 'answers'.

Don't get me wrong now. I am not buying into the notion of 'conceptual framework' as if it were something with a physical existence— an object to un/cover if we look closely enough and ask the right questions. Theories of conceptual change and constructivist pedagogies are not without their conscientious critics— those who find cracks, so to speak, when it comes to in(co)herent theory and/in practice issues. For instance, Cato (2006) explores the thesis that "constructivist pedagogy is theoretically incoherent and therefore, where it claims to differ from traditional pedagogy, practically vacuous" (p. 73).

[The math education instructor] had pictures up... I don't know where they were from... but they were teaching kids to do problems or whatever, and they gave them these problems and they did their version of it. It would be like a map and [the question was:] 'What are all the ways you could get from point A to point B?' And they had to explain their answers using pictures and words and somebody wrote— like there were four different routes you could take if you followed these roads— and one little guy said 'You could run, you could skip, you could...' (laugh). And I thought, **'that's my kind of answer!'** Like, that's not wrong. And somebody said **'how do you mark that?'** and I don't know. Because I don't think I'd want to tell the little guy it was wrong. Maybe it was the wording of the problem, too. I was kind of proud of him. (Ursula, January 2000)

After all, light **is** emitted in all directions. It is only because we physically limit or constrict light's path to direct it where we want it to go…

tunnel vision *noun*
1. Vision in which the visual field is severely constricted, as from within a tunnel looking out.
2. An extremely narrow point of view; narrow-mindedness. (AHDEL, 1992)

or, representationally speaking, we reduce light's movement to an exercise in ray tracing where we geometrically trace one single light ray through various obstacles (but with very clear 'rules' for what happens at the surfaces of each obstacle). Either way, departing from an image of the rectilinear propagation of knowledge and embracing the notion of multi-directional knowledge constructs is challenging for most teachers, as illustrated in the following conversation with Helen.

Conversing with Helen...

Kathy: My question to you is: is there a science or math that is not just about answers? Is there a way of approaching math and science so that we feel comfortable with process and other things, and not always feel that, to show our knowledge, we need to get an answer?

Helen: I think for science there is. I can see in grade 1, bringing in magnets or mirrors or something fun... they can just play with them and see what happens, see what they find out. And I'm not going to say 'well, you know, this is because mirrors are...' whatever. Just let them play with it and see what kinds of things they can learn on their own. I think that can be done for science, for a lot of areas, but for math... I don't think so. I don't know... like two plus two is four. That's all there is to it. I don't know how to make it *not* be that way. Math is a whole different thing, for me.

Kathy: I can see that, because I say I'd like to be able to focus more on explorations in math and process and problem solving... more open-ended types of problems... but then when someone asks me for an example, I'm stuck. I know it *could* exist because I know that there are ways...for instance, what are the different ways of arriving at four things? Well, you could have two alike or you could have one and then three alike; there are ways so that there is not just one answer... it doesn't mean that two plus two won't equal four anymore, but it's a whole other approach... more open ended. Have you been exploring those in your math classes?

Helen: I don't get a lot out of that class because we just get stuff and we play with it. It's not teaching me anything. I *totally* don't feel prepared to teach math right now. I can't even imagine doing math that way... way too hard to do that. My teaching is going to be a lot how I was taught, I think, in math. I can't get the new ideas about it... and I know that's bad.

Kathy: With math it is more difficult. I mean, there *are* skills, but it's unfortunate that that is the *only* way we tend to look at it...why not different groups working on the same problem and coming up with different approaches, but all equally legitimate?

Helen: That's probably a good way to teach, but I think that's just going to stress me out. Like, if I give my kids the chance to come up with their own ways of doing things, *I* probably won't understand some of the ways. Like I won't be able to see how they got there. I'll be like "...ohhhh, I don't know" (trepidation in voice), 'cause for me I'm not good at visualizing, I'm not good at abstract things, I need concrete. So, some of these kids will be coming up with things, and showing me, and I'll have *no idea* how it works. That scares me... 'cause I won't be able to understand what they're talking about.

Kathy: And that's a big thing— comfort level— you may have four or five different approaches, at least, in the class, and one of them might be what you're familiar with... and what do you do with the other ones? How does it make you feel as a teacher?

Helen: Scared (laugh)

Kathy: Well, I think recognizing it is a big thing... not a lot of pressure from day one to go out and teach completely different from how you were taught. It's experiencing that uncomfortableness— being comfortable with uncomfortable. If you're uncomfortable, it gets you doing something different... they explore on their own and then you can't help them...

Helen: That's what scares me... they'll be doing all these things and they'll want to know if their way works and I won't be able to tell them. I have *no* idea. It might work for this question but I won't be able to see how they did it, or how it might relate to something else.

Kathy: How would you *like* to handle that?

Helen: I'd like to be able to understand. I'd get them to try to explain it to me. I probably still won't get it. Um, I don't know. I have no idea.

Kathy: Have you had other teachers who've handled it differently... any teachers that just said 'Ah, no. Your way doesn't work'?

Helen: Yeah. Like *I* always did it the same way as the teacher, but we had other kids in the class that had other ways of doing things. *I* never got them, but the teacher would just look and say "well that's not the way the book does it" and [the teacher] wouldn't even really look to see *how* they did it. I think that's wrong.

Kathy: Do you think that they feel any different about it than you? Maybe it was just the route they chose to take. That bothers me... I always saw my high school teachers as being all-knowledgeable so if they *said* that this particular process wasn't the right way to do it, you know, I just agreed. When I think about it, it would have been so much nicer if they'd just said 'gee, let me look at that for a while... you may have come to a different approach', or something, but they never opened themselves up to that. Why do you think this is?

Helen: They're scared. *I'm* terrified. I'm hoping I won't get *too* much of that in grade one or two. I could *probably* handle *most* things but I'd be *terrified* not to be able to help a kid. I'd have to say 'I don't know' and that's horrible. Like for some things that's ok, but for math... to me, for math, it's not ok to say 'I don't know'. Like for science it is; for math, it's not. I don't know.

Kathy: Do you know why that is?

Helen: (sigh) I don't know. I don't know. Science, for me, doesn't have to be just about answers. Math, for me, does.

Kathy: So, there's a difference between saying you have a different kind of knowledge in science... but in math, other ways of knowing are more difficult to imagine?

Helen: I understand different ways of teaching it and different ways of learning it, like math centres. But still there's that answer at the end. Yeah, you can have fun and play with stuff to *get* to the answer but there's still the answer.

Kathy: Yeah. And every question you can think of— that you would give or have experienced yourself— has always been about one final result?

Helen: Yeah. Even if it is fun to get there, get them motivated and everything, they still have to get to the answer.

Helen has expressed clear images of the importance of answers in mathematics and her fear of not being able to help 'bring' her students to this answer unless she has everyone do it her way (often the textbook way). Several researchers have written about the concept of "funneling" in mathematics (Alrø & Skovsmose, 2002; Johansson, 2007), where the teacher-student interaction is limited to/by the process of funneling "the students through the task to the same answer as the textbook" (Johansson, 2007, p. 48).

> As I see it, the funneling pattern of interaction and the passive position of the teacher create a special context for the learning of mathematics. When students are working with tasks in the textbook, the focus is on getting the right answer—by looking at the answer key at the end of the textbook, students can be confident about their solution. (Johansson, 2007, p. 49)

Ursula expresses similar trepidation at the thought of trying to SEE mathematics in the same way her students do. She had never experienced mathematics learning using manipulative materials so she expresses concern that she will not be able to help her students:

A lot of what we're taught says that [using manipulatives] is a better way to learn [math]. But that often means that we have to re-train ourselves so... I'm just so scared of it, I don't know.

I think they *see* it a different way.
Like if they see a question,
they see that whole question a different way... I can't quite tell
what they see. Do you know what I mean?
Like I see a three-digit number in columns.
That's all I see. If you add up the little columns...
They see it as 'well, you have this many tens, so you add...'
And I just see it as, ***'carry the one.'***
I just can't picture...

(Ursula, January 2000)

Helen made a point in our conversation that she sees a clear distinction between her image of mathematics and her image of science. Helen explained that her four-month experience in an elementary science education class made quite a difference in her image of science and science teaching.

> I didn't do near enough in four months to call myself an expert
> but just my whole attitude changed.
>
> I don't know everything
> but I know I can *find out* about everything;
> there are people I can talk to, books I can read...
>
> For twelve years [my science experience] wasn't significant enough
> to totally turn me off completely...
>
> The four months have brought me back to where I'd like to be.
> (Helen, December 1999)

Even though Helen appears transformed, I believe the space is still about finding 'the' answer to demonstrate knowledge. There is plenty of encouragement these days in teacher education programs for being human as well as a knowledgeable teacher ☺. "It's ok to not have the answer, because you can always *find* it", we are told. There is not, however, much flexibility in terms of how we define what counts as knowledge. Do I have knowledge if I have answers? Yes, even if I have to find them outside of myself, but "just knowing I can find them". Do I have knowledge if I have ideas? No. Ideas cannot be pinned down as easily—it is too difficult to evaluate *the* final circled result.

In writing on the profound question of where does one *get* knowledge, Bruner (1996) states:

> If nobody in the group "knows the answer", where do you go to "find things out"? This is the leap into culture as a warehouse, a toolhouse, or whatever. There are things known by each individual (more than each realizes); more still is known by the group or is discoverable by discussion within the group; and much more still is stored somewhere else— in the "culture", say, in the heads of more knowledgeable people, in directories, books, maps, and so forth. So what do we do when we get stuck? And what are the problems we run into in getting the knowledge we need? Start answering that question and you are on the high road toward understanding what a culture is. In no time at all, some kid will begin to recognize that knowledge is power, or that it is a form of wealth, or that it is a safety net. (p. 52)

Notions of power and knowledge will be further discussed at a later point in this text but I think these words of Bruner echo some of the frustrations I am experiencing in trying to understand why we must always assume that the answer is out there somewhere and it is just a matter of finding it to prove our knowledge.

"I think I may have an answer for you, Kathy..."

The third space in a risk-taking pedagogy is about (be)coming. Perhaps we all have experienced, at one point or another in our learning, the discomfort associated with residing in this space for very long. The space and time between questions and answers, we are often taught, should be as short as possible.

Memories of Mad Minutes

Those were hard...
It always feels like a race to an answer. Before the teacher's even done giving the question, everybody's writing 'cause they gotta be done the question as quickly as possible and have that right answer. Like I'm still working on 'What is your first step here? What do you want to do first?' and people are done. And then, well, if *they're* already done, there's obviously something wrong with me so I may as well not... you know? (Ursula, January 2000)

"The enemy of reflection is the breakneck pace..."
(Bruner, 1996, p. 129)

I remember I couldn't do subtraction flash cards
I was like 'oh no, we have math.'
And Mad Minutes... oh, I hated those.
You don't remember mad minutes?
They *still* do them!
They're sheets...they have 30 addition, subtraction, multiplication,
whatever you're working on
on the paper
upside down
then they say 'ready, go'
you flip your paper
and you have one minute to finish all 30 questions
and they give you a mark out of 30

And I was a complete over-achiever in everything I did
so when I couldn't do this...

my mom bought me flash cards
and I hated those flash cards because they represented
the one thing I couldn't do.
I still hate subtraction. (Ursula, October 1999)

Ursula's negative experiences with learning mathematics were not limited to the drive for MAD MINUTE 'answers.' Although she felt greatly stressed and panicked by the pressure to find answers, she was at a loss to imagine a different way of knowing and doing mathematics.

I don't see a big reason why you can't tell kids...
why you can't admit that that's a vulnerability.
They can come up with something and you can say
'You know what? I don't know the answer to that but we'll find it'
and then find it.
I don't know how we're supposed to know everything, do everything right off the bat. And we still think like that.
Like I still think 'I'll have to have this BIG knowledge base in order to teach this' ... I don't want it to be, but I think it is. (Ursula, January 2000)

Ursula's statement above is an example of the many references in the conversations with my research participants to the unimaginable space of *never even trying to get to an answer*. Is it the only way we can demonstrate knowledge? Heaven help us, if we should stop short of 'reaching our destination'— an answer!

He was one of the better high school math teachers that I had...
but it was still all about the right answer.
But how can you teach math *without* worrying about the answers?
Like it's *important* to get the right answer in math.
They give a formula so that you can *get* from this point to this point.
So if you get lost in the middle of the formula, it *is* a problem.

I understand that you need to learn how to work through it, but I don't know how to teach that to people who are getting discouraged... (Ursula, January 2000)

'Getting lost in the middle': do you hear the negative connotation to this expression? Is it possible to reside in the middle without it representing failure? I believe it is only possible if these new spaces of pedagogical practices are taken seriously. The little boy in Ursula's comment took it seriously when he answered that he could skip or run between points A and B, between question and answer—a real image of having fun with the journey! But someone probably marked him wrong.

All of a sudden,
I'm drawn into feeling that I need to provide more details
on these s p a c e s
of NONrectilinear propagation of knowledge;
those risky, messy, in/between places with no answers in the back.
[even though I have "answers in the back" of this book]
Can I propose that these s p a c e s exist,
and are worth cracking open,
and then just *leave* them open?
It makes me wonder:
what do I know about these s p a c e s
and how to reside in them in *my* teaching and research?

How feasible is the notion that these three proposed spaces for/of/in questioning can be imag(in)ed? It seems like such a radical change to the discourse(s) of science and mathematics. Keller (1995) states that "even inside a relatively well-delineated community, different questions do get asked" and that "gradual shifts in the presuppositions defining meaningful questions and satisfying answers can and do occur" (p. 136). Unfortunately, Keller continues, "deviant questions and answers often do not get heard; the realization of change works, always, against a web of internal resistance" (p. 136). To imag(in)e how science might be different, Keller suggests that there must be a respect for the dialectic relationship between closure and openness in the practice of science and in the production of scientific knowledge. This draws me into the distinction between "scientific enterprise" imag(in)ing changes and "school science" imag(in)ing changes. I would suggest that a shift in the curriculum of the practice of *school* science is even more problematic than the practice of science of which Keller speaks.

> "The ideological commitments of modern science are deeply entrenched, and the resistance to change that they generate should not be underestimated." (Keller, 1995, p. 138)

Ask me no more questions And I'll tell you no more lies.

Reflexions...

I wonder why all the emphasis on answers.

Reflexions...

Somehow I just could not get my participants to imagine mathematics or science where we didn't have to produce answers in order to get a pat on the back, a gold star, and advancement to the next level of answer-seeking. They could imagine it (and many even lived it) in other subjects, like language arts. Why not for math and science?

Of course, I wasn't able to give them any examples of what I meant.
Not exactly anyway.

I didn't want to give them the answers to my questions in these conversations... that would be (mis)leading, wouldn't it? My research is evaluated based on some of the answers or results that I get to demonstrate that I have created new knowledge. I can't just write narratives of my participants' perceptions of knowing in mathematics and science without interpreting them and coming to some conclusions, right?

So, back to my pondering...
I wonder why all the emphasis on answers.

Whenever I asked my participants to think about a type of 'problem-solving' (yeah and why is it even called 'problem' solving asks one of my participants) where they are asked to think and talk about different strategies without ever carrying them through to an answer, they often look at me puzzled. They agree that the assessment of processes is an important part of evaluating a student's work, but they cannot imagine leaving the problem just hanging there without an answer. Maybe I was using a silly example, but it was all I could come up with.

Of course, I wasn't able to give them any other examples of what I meant.
Not exactly anyway.

I didn't want to let them know that I didn't really know what I was talking about either... that I am still searching for concrete examples of different ways of knowing mathematics and science. I *know* there are some examples and if I could only get my head around a few of them then I could prove to all the doubters out there that there *are* other ways of knowing (in) math and science. If I could just come up with some answers to their questions and doubts about other ways of knowing...

Anyway, what was I saying? Oh yeah...
I wonder why all the emphasis on answers.

HOW DOES LIGHT TRAVEL?

Part II: Particle and Wave Theories of Light

preVIEW...

In this second part of "How does light travel?" the light metaphor takes the reader into theories of wave-particle duality. This area of physics is of great metaphorical significance since it highlights ambiguous spaces where debates over ***either/or*** become transformed into discussions about ***both/and***. The light metaphor for knowing and what counts as knowing is explored in this part of the chapter by discussing the particle and/or wave properties of preservice teachers experiences of learning and knowing (in) mathematics and science.

Is light a particle or a wave? In the years since the work of Einstein, deBroglie, and others, many physicists and philosophers have tried to work out a satisfactory answer to this question. Some have suggested that the nature of light depends on the experiment...

Most physicists share a belief that the particle and wave aspects of light show complementary views of the true nature of light. They must be taken together. Either picture alone, particle or wave, is incomplete. (Zitzewitz & Neff, 1992, p. 568)

Is the dual nature of light well understood? That depends on what one means by 'understood', I guess. Bruner (1979) states that "to understand something well is to sense wherein it is simple, wherein it is an instance of a simple general case" (p. 105).

> Understanding is not properly attributed to the recitation of steps in a proof, no matter how perfectly the steps unfold from premise to conclusion, but to the "seeing" that occurs when the products of reason are re-examined— looked at— by intuition for the purpose of discerning or creating meaning. (Noddings & Shore, 1984, p. 53)

One might expect that, with/in our Western dichotomous views, the theory of the dual nature of light would be unacceptable because it is too ambiguous. The theory might become more understandable if we rationalize that scientists must have an image of **particles** of light travelling in **wave**-form. But no, this is not how the duality theory is interpreted for how light travels. The 'fact' is that sometimes light travels and behaves as a particle **and**,

"In reaching conclusions about the nature of a phenomenon like light we must rely upon observation of the effects it produces. An individual wave or particle of light cannot be seen and photographed as can large-scale waves and particles of matter. We can conclude with certainty, however, that light has a wave character from the study of interference and diffraction patterns, of its velocity, of the Doppler effect, etc. Evidence just as convincing exists that light consists of small packets of energy which are highly localized, and any one of which can communicate all its energy to a single atom or molecule... these particles of energy are known as light quanta or ***photons***." (Jenkins & White, 1976, p. 700)

at other times, it travels and behaves as a wave. How can light be **neither** particle **nor** wave but **both** particle **and** wave, yet not at the same time? I just don't understand.

On Understanding...

"... that feeling of excitement, of anticipation, of certainty is in itself an indication that the quest for meaning and understanding is part of the intuitive function." (Noddings & Shore, 1984, p. 53)

"If anybody says [s]he can think about quantum problems without getting giddy, that only shows [s]he has not understood the first thing about them." (Neils Bohr, as quoted in Zitzewitz & Neff, 1992, p. 558)

science to me
was always **way out there**

it wasn't something that I was a part of...
it wasn't something you could necessarily even ever understand...

it was something that you study
and you try to get the right answers.
(Preservice teacher, pilot study April 1999)

I think in physics, I found it very much to be math-based. For me it was formulas, and then you just take the numbers and sub them in the right place. So you don't have to understand anything to do that. You just do the action. That's not hard for me. I can fill in the same thing over and over again, but I think that teachers used to think that meant you *understood* what was going on. (Evelyn, December 2000)

Metonymic Space(s) of Doubling

Are you in or are you
out? Should I stay or
should I go? Did she
pass or did she fail?
With each choice, I
rapidly lose grey
matter, tossed to
one side OR the other,
asked to horizontally
to-**OR**-fro myself
with no (re)siding in
between. Time only to
dispel of ambiguity and
tension and let gravity
take care of any desire
(to float). Always mov-
ing here **OR** there.
Never living the **AND**.
The AND/OR space(s)
are for wafflers; for
doubters and doublers.
No (re)cognition of the
necessary CONfusion
in fusion. Look at
Einstein. How hard do
you think they laughed
when he said neither
particle NOR wave
but BOTH particle
AND wave ?
Metonymically speaking,
of course.

SAID **UNSAID**

LIGHT **DARK**

PRESENCE **ABSENCE**

GOOD AT **NOT GOOD AT**

KNOW **DON'T KNOW**

PARTICLE **WAVE**

It appears from my conversations that my research participants predominantly experienced the particle theory of knowledge construction in their learning, and much more frequently in mathematics and science than in any other subject. They did not talk about experiencing knowledge construction as a wave, with each section connecting almost seamlessly to the next. In fact, it is almost as if the two worlds (or paradigms) of knowledge travelling as particles and knowledge travelling as waves were **incommensurable**.

Reflexions...

What am I trying to say here by using words like particle theory, wave theory, and incommensurable when writing about my participants' experiences of mathematics and science learning? When I say incommensurable (like Kuhn, 1996, p. 103), I do not mean that the paradigms cannot be compared but that they cannot be compared on a one to one correspondence, point to point. Is it possible that scientists, in working with knowledge as coherent and wave-like, have attempted to make things easier to understand and, in doing so, have reduced school science to little packets or bundles of knowledge that seem disconnected from one another? Are we speaking in the language of science waves but only learning science particles (like the partic(ip)les of language)? It is unlikely that we ever really identify with the language of science because, although we *use* the language of scientists, we are not immersed in it like they are— reading and writing the literature and living the connections. In school science, we pretend to enculturate, without a clear picture of the culture. So, in a sense, the paradigms change in moving from science to school science, from scientists to students of science; since these two cultures operate in different worlds, knowledge cannot be easily compared between worlds. The two worlds are incommensurable, so to speak. And in the attempted transfer, much is lost. We lose many of the **wave**-like properties...

periodicity, regularity, punctuality, rhythm, steadiness, evenness, uniformity
timing, phasing, serialization, continuity
alternation, at every turn, by turns
reciprocity, correlation
tidal flow, ebb and flow, wave movement, tidal movement, fluctuation
to-and-fro movement, pendulum movement

and see mainly the **particle**-like properties

small thing, miniature, microcosm
particle, atom, dot, point, pinpoint
dab, spot, fleck, speck, minim, mote, smut
grain, granule, seed, crumb, powder, drop, droplet, driblet
thread, wisp, shred, rag, tatter, **fragment**, **piece**

As with the duality theory of light, we need both particle and wave properties of knowledge to form a complete picture. So, why do school science learners express much more direct experience with only the particle nature of knowledge?

I am trying to understand more about why I am drawn to using the word incommensurable to talk about the relationship between a scientist's world and the world of a student of science. Perhaps we try to simulate too much of a one-to-one transfer, especially in terms of the language. Bruner's (1996) notion of the spiral curriculum embodies the concept of readiness: "… the idea that the shape of a curriculum be conceived of as a spiral, beginning with an intuitive depiction of the domain of knowledge, circling back to represent the domain more powerfully or formally as needed" (p. xii). He also states that "… domains of knowledge are *made*, not *found*: they can be constructed simply or complexly, abstractly or concretely" (p. 119). In attempting to 'transfer' the knowledge of science to school science curriculum, I think the notion of the spiral is lost, especially with respect to language. I think teachers are always feeling pressured to use language as if the concept did not exist without the language used to describe it. When teachers make an effort to present concepts in a very orderly manner with precise and clear definitions, it is possible that students are left *only* with the emphasis on these science definitions / particles as being *the* important knowledge. In an effort to simplify curriculum, knowledge has been shattered into bits and pieces— bite-sized science morsels or bundles of knowledge that illustrate only particle properties while leaving the wave nature of knowledge to 'those who can better understand it'. Ironic, isn't it? Those who have been travelling the wave for a while, and who may possibly be the only ones who can understand science bites in isolation, are often the ones who (when involved in curriculum development) think it would be simpler to chop up the knowledge wave and serve it a little at a time, in bite-size, decontextualized pieces.

The disjointed pieces of science knowledge have been taught in such a way that my participants were often left guessing at what the entire picture might look like.

I'd like to see a complete picture of things
that's how I understand it
but my knowledge of science isn't like that…

it's like … like the game show, *Ask the Competition*… do you know that one?
you take little squares off this big picture
and so you see   a little part here
and   a little part here
and   a little part here
and underneath it all would be a *rebus*
and whoever was the first to figure out what it was…

but *my* problem is
I see   this little part here
and   this little part here
and I have *no idea*

and even if somebody were to

uncover the whole picture for me
I'd still think: well, that's a *rebus*... that's not anything I *understand*
and I'd just put it away

I don't see the connections in science clearly
I know that they exist
and I know other people can see them

(Evelyn, December 1999)

Focus on...

Jean Bernard Leon Foucault (1819-1868), French physicist. Studied medicine and then became interested in experimental physics. He carried out experiments on the speed of light which were significant in adding to the discussion on the nature of light. Prior to Foucault's experiments, Newton and his followers believed light to be made up of small particles emitted by every light source. In addition, there was another camp led by the Dutch physicist Huygens believing that light was composed of waves. According to Newton's particle (or, corpuscular) theory, light should travel faster in an optically dense medium like water than in a less dense medium like air. Huygens' wave theory predicted exactly the opposite of light's speed travelling through different media. In 1850, Foucault used an experimental set-up with a rotating mirror and a long tube containing water to measure the speed of light in water. Foucault found the speed of light in water to be less than the speed of light in air. His result was considered to be a very strong confirmation of the wave nature of light. (Jenkins & White, 1976)

In writing about a particle theory of science knowledge, one is not only tempted to make connections to photons of light but also to the abstract science concept of an atom. As a metaphor for the learning of science, the atom is very closely associated with the metaphor of light, in that atoms of knowledge are isolated bits of knowing that appear unique and unrelated to other atoms of knowledge (especially if they are contained in different textbook chapters). In the world of science concepts, however, atoms by themselves are simply parts of a whole. They are concepts in need of context. That context is what gives life and meaning to our knowledge of atomic theory. In isolation, the concepts are merely pieces.

Atoms and molecules are simply particular characteristics of this cosmic fabric...space to Einstein was like a painting, filled with itself. Particular points of the painting, like atoms, contribute to the total experience. When they are removed from the larger process, they take on a different meaning. (Kincheloe et al., 1999, p. 67)

Two atoms were sitting in a restaurant. When they left, the first said "Wait, I have to go back! I left an electron behind!" "Are you sure?" asked the other. "Oh, yes" said the first atom, "I'm **positive**!"

on seeing the unseen

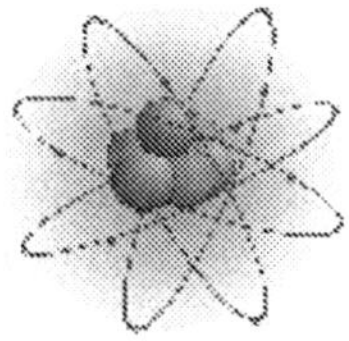

i have a picture

(what goes through your mind? not much)

made of pieces

(what purpose does this serve for you? not sure)

with no whole istic view

(do you want to teach this? no way)

fragmented memories flicker

piece

by

(unnamed) **piece made visible**

oxygen hydrogen and the iron y

to (not) **see these atoms of knowledge**

anATOMize my knowing

One of the most challenging aspects of the conversations with my participants relates to ideas of 'understanding'. As Noddings and Shore (1984) state: "Understanding, like explanation and measurement, properly varies in depth and precision with the uses for which we seek it" (p. 148). Shortly after a workshop on the atomic model in a science methods class (as discussed in the introduction to this book), I conversed with two of my participants about the workshop and its meaning for them. The concept of 'understanding' is challenging for me in my research because participant quotes like the following are both enlightening and full of ambiguity for me. Again, these 'pieces' of conversation, representing 'pieces' of my research, speak to me about the 'pieces' of knowledge that seems very characteristic of the science education experiences of my participants.

for me it [the atom workshop] **was the easiest one**
only because I remembered the most about it

the atom... I remember that *right* out of grade 12 chemistry...

I know it *should* be more difficult
because it's more of an abstract thing
(or that's what we were *told* anyway)

but for me it was easier...

I remember drawing the different molecules,
the shells

I remember doing that for everything
in the table of elements...

(well, not *everything*, but...)

I think it's because we did it over and over again
and we just kept doing it

(Elsie, October 1999)

chemistry,

I have no understanding of whatsoever,

but I could do well in the class

because it was just...

drawing bonds
I could draw the bonds
but I don't understand what 'bonds' means
I don't understand what the atom is...
it just never clicked

but I could still do well in the class
because I could still
use the formula

it's hard to teach something
not straight out of the textbook
if you don't understand *why* you're learning it
if you don't have the understanding behind it

(Evelyn, December 1999)

A very realistic analogy to the missing meaning in decontextualized pieces of science is presented by Kincheloe et al. (1999) through an explanation and comparison of photography and holographic photography.

> In photography a lens is used to focus light from an entity, so that each part of the entity is reproduced in a section of the photographic plate. In holographic photography, however, a laser light shines through a half-silvered mirror. When the beams unite at the photographic plate they produce a pattern to the viewer that produces the sensation of viewing a three-dimensional object. Contained on every portion of the holographic negative is the whole image that was photographed. (Kincheloe et al., 1999, p. 67)

I cherish this idealized image of holographic learning and knowing. It does not, however, emerge in my research conversations. In fact, not only is the three-dimensional 'life-like' nature of knowing missing, but even a regular photographic image seems difficult to achieve. For my participants, the 'picture' of a puzzle came to mind much more often.

Math is easy.

It's a game. It's a puzzle.
[Like science], math is yes or no.
But I *know* I'm going to get the right answer.
Math to me seems like sort of a closed box.

This is math, here it is, these are the boundaries and it stops right here.

Whereas science seems to be so much bigger and broader and spread out than math...
And I realize that if I were any kind of mathematician I would know that math doesn't just stop here.

(Evelyn, December 1999)

> For the mathematician's job is not pure puzzle mongering. It is to find the deepest properties of puzzles so that s[he] may recognize that a particular puzzle is an examplar… of a family of puzzles. S[he] is also a student of the kinship that exists among families of puzzles. (Bruner, 1979, p. 98)

The idea of a(n) (un)finished puzzle is a common metaphor for my participants' description of their knowledge in/of mathematics and science. When asked about how much knowledge is enough to feel comfortable, they have no idea and even suggest that they would not recognize 'enough' when/if they got there. My understanding is that, as knowledge accumulates in bits and pieces, it seems completely overwhelming to these preservice teachers. Bruner (1979) attempts to assure everyone that knowledge 'accumulation' need not worry us, as the accumulation is associated with relationship.

> In content, positive knowledge is increasing at a rate that is alarming when considered in terms of what one man [sic] can know in a lifetime. But fortunately, as the bulk of knowledge increases, the organizing structures that support it also grow. So that if there is ever more knowledge, it may indeed be the case that it is ever more related: the only possible way in which individual knowledge can keep proportional pace of the surge of available knowledge is through a grasp of the relatedness of knowledge. (Bruner, 1979, p. 108)

This quote by Bruner helps me to understand why my participants feel panic-stricken at the "bulk" of knowledge that they perceive is unknown to them. They do not have a sense of the organizing structures of which Bruner speaks that support this bulk by emphasizing and playing on the relatedness. To them, perhaps, mathematics and science knowledge seems unlimited, unbound, and unrelated; yet 'out there' and known by the experts. And this is scary for anyone, let alone a teacher who feels the pressure to *know all.*

Outlined in the theory of the dual nature of light is the belief that some phenomena can best be explained by the particle theory and other phenomena can best be explained by the wave theory of light (e.g., interference of light), leading to the integrated quantum concept called the **photon**.

> **pho·ton** (fo¹tòn´) *noun*
> 1. The quantum of electromagnetic energy, generally regarded as a discrete particle having zero mass, no electric charge, and an indefinitely long lifetime.
> 2. A unit of retinal illumination, equal to the amount of light that reaches the retina through 1 square millimeter of pupil area from a surface having a brightness of 1 candela per square meter.
>
> — **pho·ton¹ic** *adjective* (AHDEL, 1992)

The point I am making is this: if one does not 'see' the interrelatedness among knowledge bits—bundles of photons— then one sees the accumulation of knowledge as overwhelming. Where is the desire to learn more if **more** simply implies more "stuff you have to know" and, at the same time, more implies **less** in terms of connections, relevance, and inter-relatedness?

My desire is not to know more by heart
unless the meaning of knowing by heart changes...

Writing about knowledge connections and inter-relatedness (and lack thereof) calls to mind the optics metaphor of the interference of light. When waves of light are in phase with one another, they will reinforce and build on one another to create a wave of much greater amplitude than that of either one individually before the interference. This is called **constructive interference**.

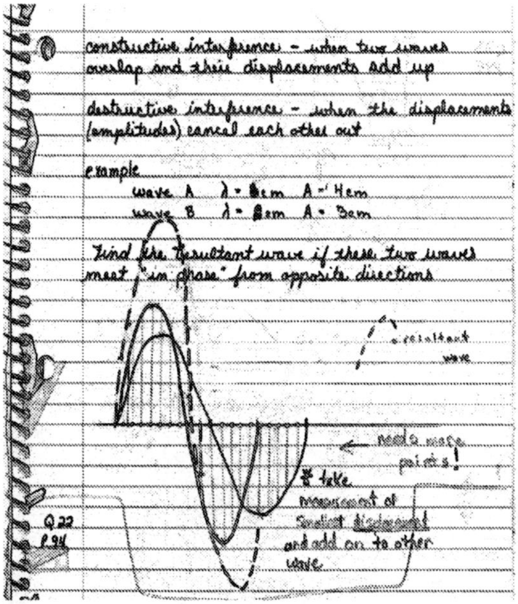

Kathy's grade 11 physics notes

When the waves that meet are out of phase, then the waves are said to interfere **destructively** and the resultant wave will not reflect a 'reinforcement' or strengthening of either wave individually before the interference; instead, the resultant wave will be less in amplitude. According to Bruner (1996), while knowledge amplitude or content may be increasing, it is doing so in a sort of 'in-phase' manner, building on and emphasizing relatedness. It is this sort of interfering or interacting with knowledge that is served well by the metaphor of constructive interference of light. It is, however, only imaginable if we experience the wave behaviour, ***as well as*** the particle behaviour, of knowledge. This metaphor is further supported by the literature on constructivism, which speaks to the notion of waves of knowledge being 'in phase' with each other.

It was apparent that the experience of the 'particle theory of knowledge', while common in mathematics and science classes, was not as prevalent in other subject areas. This realization often directed our conversations toward perceptions of subject disciplinary boundaries and subject status, which will be discussed in the next chapter on 'what happens when light hits different surfaces'.

RESPONSibilities

I've read this [section] twice and I like the substance of what I am reading but I wonder if Kathy isn't a bit of a "particle" thinker whereas perhaps I'm a "wave" reader! Somehow I become distracted by the bits and pieces of this and that. I have a visual condition known as 'macular degeneration' and as a result I see little bits of 'stuff' floating through my vision.. I blink and the stuff moves out of the way so I can see more clearly. This writing style has, to my mind, a lot of floaters in the text. Clever? Yes! Necessary? I'm not so sure!

CHAPTER TWO

HOW DOES LIGHT TRAVEL?

Part III: Formation of Shadows

preVIEW...

In this third part of "How does light travel?", the metaphor of shadows is used to critically explore the significance of particular school experiences from the perspectives of elementary preservice teachers. Through the voices of preservice teachers, this part directs one's attention to *objects* that cast shadows on the learning and knowing of mathematics and science. Three such shadow-casting objects discussed are *textbooks*, *teachers*, and *gendered ideology*. This part of the chapter seeks to critically illuminate the problematic nature of what it means to know (in) mathematics and science by examining the shadowy texture of unquestioned ideologies, including images of knowing 'beyond the **shadow** of a doubt', while often engaged in mere classroom '**shadow**boxing'.

(Bulla, 1994, p. 3)

"The fact that objects can be made to cast fairly sharp shadows may be considered a good demonstration of [the rectilinear propagation of light]."
(Jenkins & White, 1976, p. 4)

shad·ow

shad·ow (shàd'o) *noun*

1. An area that is not or is only **partially irradiated or illuminated** because of the interception of radiation by an opaque object between the area and the source of radiation.
2. The rough image cast by an object **blocking rays** of illumination.
3. An **imperfect imitation** or copy.
4. Shadows. The darkness following sunset.

5. **a.** A **feeling of gloom** or unhappiness. **b.** A cause of gloom or unhappiness.

6. **An insignificant portion or amount; a trace: *beyond a shadow of a doubt*.**

[Middle English, from Old English *sceaduwe*, oblique case of *sceadu*, shade, shadow.] (AHDEL, 1992)

What is known about shadows? They are ever-changing… dynamic, not static. They do not have a consistent darkness, as some may believe, but there is both umbra and penumbra contributing to the shadow TEXTure. We take our shadow for granted. Like a hegemonic aspect of our being, we seldom pay close enough attention to it, nor really question why it is called our dark side, and never really feel that we have any control in its formation.

> You, who are blessed with shade as well as light,
> You, who are gifted with two eyes, endowed with a knowledge of perspective, and charmed with the enjoyment of various colours,
> You who can actually see an angle, and contemplate circumference of a circle in the happy region of Three Dimensions –
> how shall I make clear to you the extreme difficulty which we in Flatland experience in recognizing one another's configuration?
> (Abbott, 1984, p. 54)

Were it not for the 'shady' parts, we would not be able to distinguish one object from another. The many wonders (and rules!) of how light is reflected, refracted, and absorbed provide us with all the knowledge we need to understand shadows… or do they? As Abbott (1984) states, we are "endowed with a knowledge of perspective" (p. 54). Without the shade and shadows of our third dimension, everything would look the same… Imagine living in Flatland!

(Photo: K.T. Nolan)

The Sundial

Within our garden close a sundial stood
With sturdy pillar fashioned from grey stone,
Its use a marvel to the neighbourhood,
Its aged face with lichen overgrown;
Yet, when a shadow cast by the sun's rays
Was slanted from the greening metal style
The time was clearly marked to our amaze,
No clock was more exact than the old dial.

But only when the shadow fell, could we
Observe the time upon the solar clock –
A truth as ancient as eternity,
A verity which eons do not mock –
Without the dark how understand the light?
Without the shadow how could life be bright?
(Dorey, 1949, pp. 1-2)

SHADOWS: THE METAPHOR

There are three types of materials, classified according to whether/how they 'transmit' light:

Transparent... material transmitting light without distorting direction of rays
Translucent... material transmitting light but distorting its path
Opaque... material that does not transmit light
(Zitzewitz & Neff, 1992, pp. 725-727)

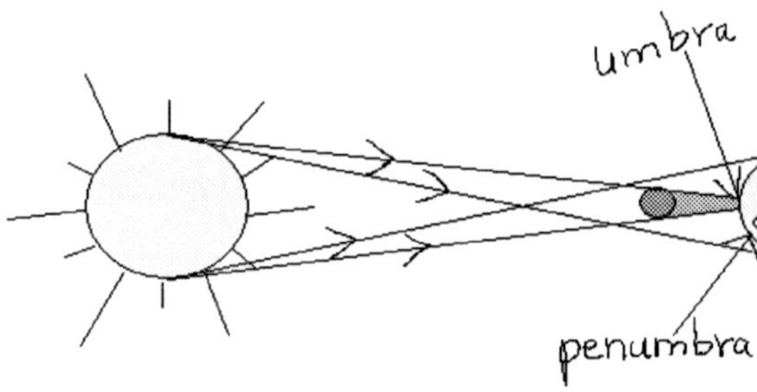

The metaphor of shadows is used to critically explore the significance of particular school experiences from the perspectives of elementary preservice teachers. The voices of the preservice teachers direct one's attention to *objects* that seem to be shadow casting in the learning and knowing of mathematics and science. Due to the property of rectilinear propagation of light, if an object does not transmit or reflect light then it will cast shadows. One of the interesting things about shadows is that the objects casting them do not, and cannot, block *all* light. In fact, anyone who has found themselves to be in a dark room will attest to the fact that darkness is an illusion, and that within seconds it becomes clear that we are rarely in the *presence* of complete *absence* of light. Each object in the room slowly comes into a shade of grey focus.

In the learning of mathematics and science, there are several objects that cast shadows on our knowing. These objects do not make it impossible for us to learn since they do not block out all of the accessible views on knowledge, but it appears that they have become significantly large and opaque in their influence on/in school mathematics and science. Three such *shadow casting objects* that will be discussed in this chapter are ***textbooks***, ***teachers***, and ***gendered ideology***. The influences of textbooks are presented from two main perspectives: 1) how they are *written*— often emphasizing the direct transmission of 'right' and 'true' facts in a language of its own decided abstraction and, 2) how they are *used*— often involving rigid adherence, with an emphasis on isolated chapters as isolated particles of knowledge. The discussion then naturally weaves into the influences of teachers, where issues of knowledge and identity are discussed with reference to how teachers are imag(in)ed to know *beyond the* ***shadow*** *of a doubt*, while often engaged in mere classroom ***shadow-boxing***. Following a greater awareness of textbook practices and identity construction issues in the classroom, the focus then shifts to the importance of the means by which teachers transmit or transfer the *effects* of these power/knowledge issues to students, especially in terms of gendered messages and expectations.

In presence there is absence

The shadow image is used for several reasons. Firstly, it is known that an object is capable of casting a shadow that is much larger than the object itself. As a teacher, it is difficult to know the impact of a statement that has been interpreted in ways other than intended due to language, tone, or any number of other signs used in interpretation. It is important to acknowledge that language is not absolute; we only *think what we think* and *know what we know* through our personal interpretation and sense-making of events and language. Meaning is always in flux and is continually (re)shaped with/in language. Secondly, the shadow metaphor is used to represent ideology that has become hegemonic. In this part of the chapter, I write about the shadows between words and actions that, in many cases, speak louder than the words and actions themselves. And finally, the shadow metaphor is used for the simple reason that the meaning of the word shadow expands to include a general feeling or cause of gloom or unhappiness, which is often the image that comes to mind when these preservice teacher participants speak about their experiences of school mathematics and science.

TEXTBOOK SHADOWS

To some extent, ideas of textbook learning have already been discussed in the rectilinear propagation section, relating to issues of how textbooks are used. In this section, I consider issues of language in mathematics and science. As I will discuss, a student's perception of what is important and what counts as knowledge is often closely tied to her/his encounters with the language and presentation of textbooks. In terms of presentation, a standard text format often involves the presentation of unknown terms in **boldface** font and then defined at the end of the chapter or in a textbook glossary. Visual materials, such as photos, diagrams, maps, etc., are often set aside from the verbal materials, such that the 'text' is seen as the real stuff of knowledge.

> When illustrative and tabular materials are sharply separated from text, it is easy to skip over them, saying to one's self, "I'll come back later and look at this." Therefore, if the materials play a serious pedagogical role and are not included for mere window dressing, they probably should be incorporated in the text itself in a way that makes their scrutiny unavoidable. (Noddings & Shore, 1984, p. 145)

Noddings and Shore (1984) believe it is important "… to embed visual materials in the verbal materials in such a way that they—along with the words—contribute to the creation of representations" (p. 145).

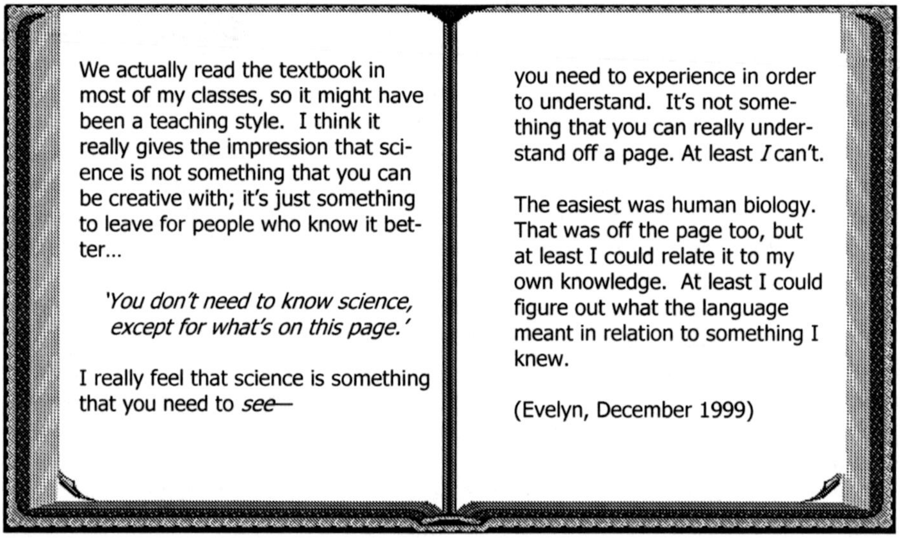

In this discussion of textbook science, the intent is not to perform a textual analysis, but instead to explore its *shadows*. In research on the use of language in school science, Sutton (1996) expresses a belief that in "recent years school science has often been conducted as if it were a study of Nature directly, rather than a study of what people have thought and said about Nature" (p. 15). He claims that textbooks perpetuate common associations of science with established facts and inflexible rules. Noddings and Shore (1984) remind us that "there is little reason to suppose that precise language used in instruction contributes much to the understanding of students. On the contrary, it would seem that such language is a product of understanding and not a precursor to it" (p. 120).

Lemke (1990) notes: "The style of scientific language is created partly by grammar, partly by choice of words, partly by its use of idioms and metaphors, and largely by its avoidance of many stylistic devices that are freely used in other kinds of language" (Lemke, 1990, p. 130). Instead of encouraging the writing and (re)writing of the rules, patterns, practices, and categories of scientific language in and through our lived experiences, we are taught to decontextualize and dehumanize science. In other words, we learn context-free rules, patterns, practices, and categories, and then later try to fit our lived experience into the constructed framework. I am not suggesting that the rules, patterns, practices, and categories do not have the potential to further our understanding of the lived world, only that we should be more open to examining the legitimization of rules, patterns, practices, and categories through lived experience and not the other way around. We should be critical of ***a*** worldview that purports to be ***the*** worldview. Bauer (1992) states that "there is a difference between knowledge about the world and knowledge of what scientists believe about the world" (p. 11).

They gave you a handout, you memorize it quick the night before and you go spit that out and forget it the next day. I'm very good at [definitions]. If I *had* to have a science test, I'd want it to be that... 'cause I can rattle it off for you. If you look at that Bloom's taxonomy that they teach us, if you look at science— at the stuff *I* learned in science— it's all right at the bottom, it's just knowledge, just the knowledge level. I never applied any of it.

I think we need to ask children to apply it somehow, but I don't think we can do that until we make it ok and find another way to teach science so that it's possible... 'cause I don't think I had the support to find a way to apply it. Like I wasn't getting anything but facts from a book, so how would you know how to...?

(Ursula, January 2000)

Teaching school science as formal textbook science is a seriously misleading (and, in fact, often missing) representation of the construction of scientific knowledge. Bauer (1992) states that in high school textbooks "one learns what is known with considerable certainty, in compressed form, so that one can remain quite ignorant of the fits and starts and false trials and silly beliefs that were amply displayed when the textbook knowledge was first being gained by human beings" (p. 10). The goal of textbook science seems to be to portray 'normal' science as objective and unchanging, which, of course, it is not.

> [As] pedagogic tools for the perpetuation of normal science, [textbooks] have to be rewritten in whole or in part whenever the language, problem-structure, or standards of normal science change. In short, they have to be rewritten in the aftermath of each scientific revolution, and, once rewritten, they inevita-

bly disguise not only the role but the very existence of the revolutions that produced them." (Kuhn, as quoted in Threadgold, 1997, pp. 17-18).

Inside Research

Kuhn (1996) explains that transformations or revolutions of scientific paradigms are common in the development of scientific theories. "Today's physics textbooks tell the student that light is photons, i.e., quantum-mechanical entities that exhibit some characteristics of waves and some of particles... That characterization of light is, however, scarcely half a century old. Before it was developed by Planck, Einstein, and others early in this century, physics texts taught that light was transverse wave motion... ", and before that, in the eighteenth century, Newton's theories "taught that light was material corpuscles" (Kuhn, 1996, p.12).

A scientific paradigm, once 'formulated', is generally accepted by the scientific community until/unless another one comes along which proves itself more fruitful, and then it replaces the previous one. Kuhn remarks, however, that this is not *always* the way things proceeded. "No period between remote antiquity and the end of the seventeenth century exhibited a single generally accepted view about the nature of light. Instead there were a number of competing schools and subschools. ... One group took light to be particles emanating from material bodies; for another it was a modification of the medium that intervened between the body and the eye; still another explained light in terms of an interaction of the medium with an emanation from the eye" (p. 12). Kuhn goes on to explain that each of these schools would illustrate their theory in terms of what it was best at explaining and how it was related to a particular metaphysic. Kuhn states that based on an "intertwined theoretical and methodological belief that permits selection, evaluation, and criticism" (p. 17), it is no wonder that the similar phenomena would be described and interpreted differently when informed by different metaphysical, personal, and historical perspectives. "What is surprising, and perhaps also unique in its degree to the fields we call science, is that such initial divergences should ever largely disappear." (Kuhn, p. 17)

This, I maintain, is one of the critical weaknesses of curriculum, and textbooks in particular. Perhaps science would be less "mystical" to students if we portrayed it more as a human endeavour, emphasizing both its false starts and its divergent theories and alternative perspectives.

In presence there is absence

Threadgold (1997) is perceptive in highlighting that "what this kind of writing seems to accomplish is the erasure of the intertextual histories of scientific texts, the loss of the trace of the previous paradigm, the silencing of the dialogue between paradigms" (p. 18).

Since the Victorian era, science writing has omitted the mistakes and false starts that scientists experience, including only the 'objective' results, which are almost always reported in the passive voice. "The losses we have experienced because of our acceptance of this idea of never-failing science have been of two types: 1) the public became increasingly *unaware* of the limits of modern science and 2) contemporary science writing, both popular and professional, makes science *appear* infallible." (Golden, 1992, p. 158)

An interesting perspective on conceptions of language in science—as an interpretive system and as a labelling system—has been proposed by Clive Sutton (1996). He suggests that school science textbooks deliberately avoid any presentation of language as a 'flexible tool of thought' used in 'persuading others toward a new point of view' and instead move directly to the use of language as a label for the transmission of a fixed body of knowledge (pp. 5-6). He suggests that many students (and teachers) do not feel a connection with the language of science because of the ways in which the initial thinking and pondering stages of scientists, necessary in the construction of new scientific ideas and language, have been completely omitted from textbooks. Textbooks seldom write about the stages where scientists pondered, wondered and imagined.

> With science, they never asked you
> **'what do you think...?'** or **'how could you...?'**
> Science was always 'fill in this blank space here', right or wrong,
> half a mark here maybe, if you're lucky
> (Helen, December 1999)

So, the question is: How can we (re)place lived experiences into our stories of science knowing? How can we (re)value the thinking and pondering stages in constructing science knowledge where we say "I think it may be..." or "Imagine if ..."?

I venture to say that this close-ended nature of the language of science is an UNequiVOCAL influence on students who claim to 'know nothing' about the topic of electricity or the model of the atom before they have *received* formal instruction in these topics. The procedure, as the students have been taught to expect, is to "learn" the grammar of science (with all of its rules, patterns, practices, and categories), then try to fit (mold, shape) personal lived experiences into the framework. Think of all the silenced and voiceless young children in our society if we approached the science of language in the same way!

As long as the nature of the language of science is viewed as a means for labelling and transmitting knowledge, rather than as a vehicle for tentative expression and interpretation, I think we will continue to see the creation of many coping (rather than learning) strategies in both mathematics and science education.

With science...
the way it's portrayed is you have to learn each little thing
you have to know this and this and this...
you have to memorize this and this...
a whole bunch of facts
and details
and stuff
that you *have* to learn

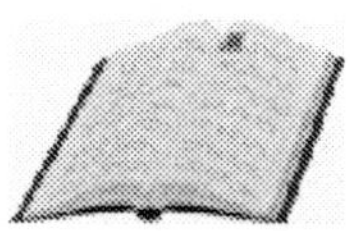

> and you *have* to be able to put those out.
>
> There's so much stuff you have to know.
>
> But with other subjects, like Phys. Ed. and Language Arts,
> as long as you can *do* it, then you're ok...
> (Helen, October 1999)

I think that in science (and electricity especially)
there are **so many terms** you hear all the time
and if you don't *know* the **definitions** too
then you don't feel you *know* about it

I *know*,
if I turn the switch   a light comes on
But maybe I don't *know* all the **terminology** behind it
And I think that's what makes me uncomfortable
about saying
    'yeah, I *know* about electricity'

'cause I *know* **watts** & **volts** & **all that stuff**
but I wouldn't be able to **describe** what each of those is **for**
or what they **do**

(Preservice teacher, pilot study, April 1999)

> We believe that one way to counter linguistic dominance is to shift *what* can be talked about, as well as the discursive form and language in which it must be spoken. In this, the ground of oppression must shift from the silent sphere of the individual as personally inadequate to the challenging speech that names the social and power relations that form the oppression— the "personal" as political. (Rockhill & Tomic, 1995, p. 213)

Hildebrand (1998) writes that "power permeates the language practices of our society, providing us with an acceptance of the allocation of roles and the distribution of privilege, as we become committed to the maintenance of the 'normal'/current social order" (p. 350). We are so wrapped up in "the image of what counts as science learning … established through an emphasis on recalling facts, concepts, and theories; solving algorithmic problems; and adopting scientists and metaphors" (Hildebrand, 1998, p. 349) that we are unable to talk about and continue to construct our own personal science knowing.

There are several coping strategies that learners can create in (non)learning. One such strategy is to make comparisons between what you know and what you think you **should know**. This strategy often leads to a (false) portrayal of indifference and disinterest in the concept being learned/taught. The learner becomes frustrated by the 'amount' of knowledge (not) possessed, and as a defense mechanism, s(he) shuts down to further learning.

Related to this coping strategy for (non)learning, is the way in which many students will confess to "just not being good at science". Like a birthmark labelling physical appearance, this claim to being incapable of understanding science appears to label one's mental capacities; a situation where 'you either have it or you don't'. These ideas are further developed in the next chapter on "what happens when light hits different surfaces?"

One of the ways to recognize and disrupt the hegemonic structure of the language of science (at work in the above coping strategies) is to struggle against its overpowering ability to manifest itself in personal failure rather than as a political and social construction. Rockhill and Tomic (1995) use the phrase "between speech and silence" (p. 213) to describe how language works to subjugate knowledge and experience.

A Moment in HERstory...

I am given a seemingly simple task:
Take a piece of paper
Turn it into a sphere
my mind turns
and in doing so dumps all of my knowledge
2-dim into 3-dim...?
(i **should** ***know*** this)
volume of a sphere...?
(i **should** be able to do ***this***)
corners into not-corners...?
(i **shouldn't** have to watch ***her*** approach)

my head pains
as I shut the door
and think about how...
tired i am
much time is left
i must look to everyone else

REFLEXIONS...

i can't believe after all this
time and reading and pondering
i finally understand the phrase "personal as political"
thoughts of
'every vote counts' and 'we the people'
were all i could fathom
now i see that...
if the oppressed & marginalized
remain oppressed & marginalized
and feeling inadequate
and self-blaming for their own conditions
then who will be strong enough to challenge...?

So, how powerful is the textbook **shadow**? It certainly appears from the conversations with my participants, and from my own personal experiences with textbook learning and teaching, that the influences of textbooks dominate over all the influences of other learning resources in the classroom. Apple (1986) states that the curriculum in most schools in the United States is defined "by one particular artifact, the standardized, grade-level-specific text in mathematics, reading, social studies, science (when it is even taught), and so on" (p. 85).

> It is estimated, for example, that 75 percent of the time elementary and secondary students are in classrooms and 90 percent of their time on homework is spent with text materials.... While the text dominates curricula at the elementary, secondary, and even college levels, very little critical attention has been paid to the ideological, political, and economic sources of its production, distribution, and reception. (Apple, 1986, p. 86)

The course syllabus for this class...
I *do* want to learn it but I look at the headings and I'm scared.
Even teaching fractions... if it's beyond cutting the circle up, I don't know.
I think it's just the words. Like all I can think of for math is chalkboard and rows of numbers. We never *did* anything else with them. We just had a textbook and we learned that...

(Ursula, January 2000)

Maybe we put too much emphasis on the *word* and the things that make it difficult. Like science words are *scary*. They're like ten feet long and don't necessarily make sense to someone who's just walking by so maybe if we just talked about the people who did science and what they were thinking and how they got to this, instead of the formula and the word and... like

> you can *put* the word there but… maybe you should know how it *happened* first. And I think that's any subject, not just science.
>
> (Ursula, January 2000)

Some readers may quickly dismiss these discussions of the language of textbooks and how they are used because they believe the textbook to be a neutral representation of scientific knowledge— even if they do exhibit language intransigence, imagination deficiencies, and failure to incorporate stories of the changing nature of science. As mentioned previously, this view of the 'neutrality' of textbooks' illustrates how the transmission of ideology can become hegemonic.

> Texts are really messages to and about the future. As part of a curriculum, they participate in no less than the organized knowledge system of society. They participate in creating what a society has recognized as legitimate and truthful. They help set the canons of truthfulness and, as such, also help recreate a major reference point for what knowledge, culture, belief, and morality really *are*. (Apple, 1993, p. 49)

Of primary influence
the teacher
knowing magically
when odd
when even
how many
it's right there
on the page
still circled
from last year's class…
same chapter same questions same teacher same students
same same same same same same same
textbook
is
was
and remains
the shadow
of knowing pursued
where you've been
where you'll go
magically following lines not edges
filling space to erase
the missing dimension
not seen on any page

TEACHER SHADOWS

Many of my participants recognized and expressed very clearly how they **did not** want to teach, but they were not so clear on how they **did** want to teach. They expressed discomfort with not being able to be the best possible teacher of science that they could be for their students, as illustrated in the following brief excerpt from a conversation with Sylvia and Harriet:

Sylvia: They'll just learn what I'm able to teach them or learn on their own, but for me to be the knower of all, that's just impossible... makes me feel almost ashamed 'cause I won't be teaching them as well in science as I know I should be... but I don't know what to do about that.

Kathy: Are you basing that 'should be' on *your* experiences... you know what you *don't* want?

Harriet: Oh, I can tell you exactly what I *don't want*, but to tell you what I *want...* Like I *want* to be their cheerleader, to lead them down this glorious pathway of science, BUT— and this is a big BUT— even after 4 months [of a science education class], I think I know how to do it to *some* extent but I think... and maybe I'm being too hard on myself. I have high expectations of myself and if I slip up then I've totally led them down the garden path and I'm leading them back to this negative aspect. Do you know what I mean?

Teachers have been important role models for my participants, in many ways—good, bad, indifferent, disinterested, enthusiastic, etc. In this section, I focus on the ways in which teachers have (un)knowingly cast shadows on the mathematics and science knowing of my participants. Two metaphorical threads can be woven into this discussion: Ideas of teachers portraying knowing 'beyond the **shadow** of a doubt' and the concept of '**shadow**boxing' in place of authentic teaching.

> I always saw my high school teachers as being all-knowledgeable, so if they said that this particular process wasn't the right way to do it, you know, I just agreed. When I think about it, it would have been so much better if they'd just said: "Gee, let me look at that for a while; you may have found a different approach", or *something*, but they never opened themselves to that. (Kathy, in conversation with Helen, December 1999)

Why is the notion of *teacher as knowledge expert* such a widely accepted image? Why is there so much at stake for teachers' identities unless they espouse knowledge beyond the **shadow** of a doubt? It seems unlikely, for professionals in such a respectable and caring profession, that they simply wish to grasp all possible opportunities to exercise their authority and power. So, it has prompted me to search for other reasons—for reasons that are more meaningful and understandable.

> Most teachers do not want to alienate their students or to make them feel like strangers when confronted with predefined, unfamiliar structures of knowledge. It is far too easy to make curriculum seem external to the search for meaning. It is far too easy to make it into an alien and alienating edifice, a kind of "Crystal Palace" of ideas. … Nevertheless, it is important to confront the notion that teaching always signifies a type of intervention into someone else's life. (Greene, 1973, p. 175)

Beyond the **Shadow** *of a Doubt*

So many issues weave into formations of teacher identities, issues that contribute to how/why teachers might feel the need to portray their knowledge(s) as being beyond the shadow of a doubt. In this section, I propose some ideas based on my own experiences and those of my participants (in their own voices) that might help to open spaces for a re-visioning and re-positioning of teacher authority.

Shadow Reason #1: Pressure

> I think the pressure from the principal, the parents, outside sources, saying:
>
> *'Where are the answers?'*
> *'What are you teaching my students?'*
>
> You have to have something to prove to others that you are teaching your students something. I think that's the pressure to have an answer or find an answer. It's an uncomfortable feeling if there isn't an answer or you don't know it or you can't find it. That concept of 'if you don't have an answer then you don't have the knowledge'... yeah, they go hand-in-hand.
>
> (Maude, April 2000)

Gordon (2006) corroborates Maude's story in her concern about the pressure to find and know answers when he asks "who can blame teacher candidates for wanting recipes when most of the conversations they hear about schooling are full of the language of accountability, measurement and testing?" (p. 17).

Shadow Reason #2: Cycle of (Non)Questioning

As a self-quote on the previous page indicates, I did not even *think* of questioning my teachers' ideas about/and knowledge, especially since they seemed unwilling to question themselves (out loud). As several of my participants commented, in order to question science and someone else's knowledge of science, one has to feel pretty confident. Some told me that they did not even know where to begin to question.

> I don't think we feel *capable* of questioning it [science].
> I don't think it would ever cross my *mind* to question it
> because I feel like, well, I don't know enough to question it.
> And in some case I guess we don't, because we only know *that* theory
> Most of the time we know *that* theory and what's good about it;
> we don't usually... we don't *always* get told why... or what might be sceptical.
>
> You have to doubt something first and if you're not given any room to doubt then how are you going to question it? I didn't question anything anybody told me until university, in education. I feel a lot more confident now in asking 'why'd you do it this way?'
>
> I think I would rather have people understand—this is maybe why I'm not a researcher—I'd rather people understand what I *do* know than pretend to know more than I *did*... like pretend this is the be-all-and-all if I'm not sure it is. (Ursula, October 1999)

Teachers will often quickly dismiss student questions of 'why'. This may not be done with the intent to frustrate students but frustration is often the outcome. I believe it is also frequently done (subconsciously) as a protection mechanism to ensure that students do not stray out of a teacher's (knowledge) comfort zone. In the following quote, Harriet replays a common, yet disturbing, interaction with her math teachers.

I'm always asking questions and I think that gets me in trouble because even my math teachers used to say
'*Don't question, just do it.*'
'But... I want to know why.'
'*Don't question, just do it.*'
No, but...'
'*No [Harriet], don't question, just do it.*'

(Harriet, December 1999)

Reflexions

The statement about not entertaining questions as being a protection mechanism is partly an interpretation I am making from my conversations with my participants. Mostly, however, it is a statement drawn from my own personal experience— I used it as a teacher myself. It is partly why I followed the text so closely and partly why I did not encourage student explorations beyond **my** comfort zone. The situation is very problematic and intertwined in ways that are difficult to articulate, but I believe that none of these "**shadow** reasons" exists in isolation. I would not have been reluctant to move outside my comfort zone if I did not perceive so much public pressure to be the expert, and I would not have perceived so much pressure to be the expert if the public had experienced a non-traditional type of mathematics and/or science classroom. I perceive things be/come even more complex than this when considering multiple shadow reasons as they interplay with each other.

During a conversation with Ursula, she expressed her frustration that, in her science education, she was never given choices on what she would like to learn and how. I asked Ursula if she had considered that her teacher did not teach in a way that gave students more choices in their learning because s(he) was afraid of where students might go. Ursula answered that she had not considered this at the time she was a student but that she had recently noticed some fear and discomfort with her pre-internship co-operating teacher:

> Now I see a lot of fear, even in pre-interning and stuff...
> I did a literature thing with [the students] about why some things were special to them and in the end they made patches for a quilt, just on construction paper... and that was *way too much* of a stray for [the teacher] from: 'I read the story, you answer the questions'. She just hated it.
>
> (Ursula, October 1999)

Shadow Reason #3: Disinterest

> I didn't understand anything in grade ten [math].
> I thought 'oh, I'm just bad at math.'
> And then I gave up.
> He'd come into class,
> He'd write something on the board that didn't make any sense.
> He wouldn't explain it.
> He wouldn't answer questions.
> Then he'd leave the class and let us work on our homework...
> That's all he did. If you asked him for help, he'd say 'well, read your text.'
> So that wasn't good.
>
> (Helen, Research Participant, November 1999)

A significant number of teachers end up teaching mathematics and/or science 'against their will'; they are not specialists and, on the surface, they appear to have little interest in these subjects areas— teaching or learning them (Ball & Bass, 2000; Kajander & Dionne, 2005). For the sake of students trying to learn mathematics and/or science— and for the sake of the teacher trying to teach it— teacher education programs must initiate critical reflection on issues of experience, attitudes, beliefs, and possible (dis)interest of preservice teachers as a way to address this shadow of disinterest that could be cast on mathematics and science classrooms.

Shadow Reason #4: No Real Experience

> I think we're hypocritical a lot of the time...
> We're taught things on such a superficial level
> and then we spew it back to somebody else.
> *We've* never done it ourselves
> but we're preaching at somebody else that *they* need to do it.
>
> If I do *one* thing, that would be it—to do it too.
> If *they're* learning something in science,
> chances are I'm going to have to learn it too
> so I can't see why they can't *know* that.
> (Ursula, October 1999)

In this quote, Ursula touches on the notion that school science is seldom reflective of the science actually practiced by scientists. One could certainly read this as a call for more "authentic school science" (Roth, 1995) experiences, but I believe it also reflects an embedded call for teachers to take more risks—requiring them to step out of their own comfort zones with respect to what they perceive they know (and feel they *should* know) as a mathematics and/or science teacher, and be willing to *experience* learning themselves.

Shadow Reason #5: Loss of (W)holistic View

In comparing teaching to a labor process, Apple (1993) refers to one consequence of a very compartmentalized curriculum as the "separation of conception from execution" (p. 121).

> When complicated jobs are broken down into atomistic elements, the person doing the job loses sight of the whole process and loses control over her or his own labor since someone outside the immediate situation now has greater control over both the planning and what is actually to go on (Apple, 1993, p. 121).

This topic has been thoroughly discussed in earlier 'compartments' of this chapter, paralleling the metaphor of the particle theory of knowledge. The particle-theory paradigm seems to have been around for so long, evident in textbooks and tests of definitions, facts and skills, that it is no wonder why the task of creating a (w)holistic view for students is so foreign.

This compartmentalized model is disturbing in the realization of what gets sacrificed in place of 'covering curriculum' and 'performing the plan'. The following conversation that I had with Evelyn talks a little about what happened in her experience of learning facts.

conversing with Evelyn...

Evelyn: Physics in grade 11 and 12, our assessment was all formulas. But I don't think I really ever understood optics or waves or… I *know* I didn't understand force and motion, but I can fill in numbers anywhere; that's not a problem. But understanding *why* it happens… well, it's not so clear.

Kathy: I'm just wondering if the teacher in that situation would *rather* assume that everybody understands it because, well, everyone is coming out with 80's and 90's after substituting into formulas, so maybe the teacher doesn't want to go where there will be problems in understanding, right?

Evelyn: I think it's a problem in all areas of the school, because if you're a primary teacher you don't want to necessarily spend a lot of time getting kids to think about *why* when, as soon as they get into high school, they're going to have to not think about why but just have to work with the formulas and the facts. And the same thing in the high schools… the teachers don't want their students to be questioning science and figuring things out on their own if they have to get high enough marks to get into *university* where they have to be an academic in a certain way in order to be able to perform introductory physics… and universities aren't going to make allowances necessarily—probably especially in the math and sciences, for people stopping and thinking about *why*… that's what you do after you finish the first two and a half years of your program… then you can stop and think about the whys, but before then you just have to get the facts

Kathy: And then, for me, it is too late; I don't want to have to go back and admit … well, I'm doing it now, aren't I? (laugh). All I knew how to do really was use the formulas. I wouldn't have understood *why* until I started teaching it and then I felt a little bit gypped in my education because it would have been kind of neat to understand that before. It sounds like a banking thing… like putting money in a bank account…just put this formula in and this formula; keep *banking* all of these facts until you have enough to go to high school and then until you have enough to go to university. Does the banking model work so well?

Evelyn: It doesn't create learners. I don't think for the most part the school systems are designed for learners anyway. They're more designed for people sitting quietly and answering true or false questions.

Kathy: It takes a very *daring* individual to break out of that as a teacher, because you're going to come up against things you're not comfortable with, things you don't know…allowing students to construct their knowledge… they're not all going to all construct it the way you want them to. (laugh).

Evelyn: I don't think there's much support for that… if you're doing things like learning centres in math… parents and administers thinking these kids are playing all day… 'what are they learning? WHAT ARE THEY GOING TO DO WHEN THEY GET TO HIGH SCHOOL? THEY'RE NOT GOING TO BE PLAYING GAMES!' and the teacher has to say 'Well, actually you're right, they're not.'

Shadow*boxing*

Quote of the day...

I don't think for the most part the school systems are designed for learners anyway. They're more designed for people sitting quietly and answering true or false questions. (Evelyn, December 1999)

As a teacher (and as a teacher reflecting back now on my teaching), I often got the impression that the only person I was engaging in my mathematics and science classes was me— a sort of **shadow**boxing.

shad·ow·box

shad·ow·box (shàd¹o-bòks´) *verb*, *intransitive*
sha·dow·boxed, sha·dow·box·ing, sha·dow·box·es

To spar with an imaginary opponent, as for exercise or training purposes.
— shad¹ow·box´ing *noun* (AHDEL, 1992)

my teacher was very knowledgeable about it [physics]
he knew the ins and outs, backwards and forwards...
he almost knew it too well to teach it to us

and so in some respects
he was teaching to himself
but he didn't get it... down to our level
where we could grasp on to it

you know, he's a very knowledgeable man
and a great teacher...
but it just didn't come across that way

most of the time I remember coming out of there
just being completely confused...
not knowing anything

(Preservice teacher, pilot study, April 1999)

Planning...
curriculum conceptualizing, cross-
referencing carefully, clipping comics,
clever clichés, and chronological charts;
creating collages, concerned with comprehension.

Teaching...
circuit conductivity, chemical compounds,
close correlation, capsizing cantilevers;
convex/concave corrections; cotangent/cosecant
continuity, centripetal/centrifugal complexities.

Thinking...
chapter churning, clockwork
cognition, catalytic catch-up, clairvoyant
compassion; coaching, coercing, commending,
concentrating, caring, confidence, clarity, conscientious.

Observing...
cacophonous chorus of capricious challenge
creating a calamitous campaign, commonly confronted as...

yawning indifference.

With chem[istry], you need to know the basics...
In grade 12, I had to quit...
It was him in a way but it was also me...I was not getting it.
I even went for help and he'd be explaining it to me and it was like...
you know on the *Peanuts* cartoon 'waaa, waaa, waaa, waaa'?
He was the *Peanuts* teacher and he'd ask 'do you understand?'
and I'd say 'yeah'
just so I could leave,
just so I could get out of this nonsense.
My friend ended up doing my homework for me.
(Harriet, December 1999)

GENDERED IDEOLOGY SHADOWS

In this book, I have written considerably on the language of science and mathematics, especially as portrayed in textbooks, as being an UNequiVOCAL influence on *what it means to know*, as perceived by my preservice teacher participants. What also requires attention are other forms of language in the classroom that seem fairly opaque, at least apparently so to my preservice teacher participants before our reflective conversations. As students, these preservice teachers did not often see through/beyond their *teachers'* perceptions of them as mathematics and science knowers. Upon reflection on the experiences of my participants, many gendered messages came to the fore. The messages were both explicitly articulated and unarticulated, but gave students clear messages regarding their teachers' perceptions of them as mathematics and science knowers.

[See **(En)gendering Gender** on the next page]

In grade ten, when I was told—basically the guy was a  —
and he was telling me that
the girls shouldn't be doing dissections
(Roberta, December 1999)

When asked if she remembered any experiences that seemed to send gendered messages, Sylvia responded:

I don't know,
you think about it
and then you stop thinking about it...
I mean, what can I do about it anyway?

(En)gendering Gender

In grades seven and eight we had male science teachers. And it was the usual
'if the girls don't want to do the dissections, the boys are going to have to do it for them.'
Not much encouragement that way.
And in high school all my instructors were male.
I took all the sciences, algebra, geo-trig, and calculus.
And I think there was a divide between what was expected of the boys and of the girls.
Even the examples used in class.
Our chemistry teacher...
told us that the only real thing that the girls in the class could get from chemistry would be the chemical equation for diamond.
(Evelyn, October 1999)

our grade 8 teacher...
We did experiments but he was always gung ho with the boys.
The boys always got to be the *demonstrators*, and...
You know, after a while you don't even raise your hand anymore, 'cause you know you're not going to get picked.
Like I *knew* with him.
You were *not* picked with him, *ever.*
The boys were.
(Harriet, December 1999)

I don't know if this has anything to *do* with your research but
I *loved* our female math teacher
but I always felt like she was teaching to the *male* students in the class.
Her favourites were always the males. And a lot of other people noticed that too.
It was mostly with help—
if something *wasn't* explained as well as it could be
She would offer them more help than she'd offer me.
That was basically the big thing that I noticed.
(Elsie, October 1999)

You know, the guys in our school were like
'oh, the girls can't do physics, the girls can't do math.'
It wasn't so much that they *said* 'you can't do it',
but they'd say
'why don't you take something else instead?
I'd ask guys 'can I do physics, do you think I can do it?' and one guy answered,
'well no, probably not.'
My boyfriend's really smart...*he's* in physics, so he's really smart. He doesn't say that I could do it either. I asked him and he was like
'oh, I don't know.'
If he would have said 'sure, you can do it', I might have.
(Helen, November 1999)

Reflexions

These experiences of "gendered ideology" are significant in my research, and yet they are also the most problematic aspect for me to write about. Even though I use the word 'ideology', I think 'hegemony' is more appropriate because of the extent to which certain fundamental issues are not even acknowledged, let alone critically questioned. By this I mean that hegemony exists more through what is unsaid and not done, than by what is said and done. I think **shadow** is a perfect word for what I am trying to name in this research. In a world that is awakened to ideas of gender equity issues—to ideas of "female-friendly" (Rosser, 1997) and "gender-inclusive" (Barton, 1998) science— one often finds that the injustices are not so much in the words and actions, as in the **shadows** of the words and actions. These **shadows** are in need of deconstruction, even as they dance and dodge with/in the words of this research text.

> It's so dark here where I am searching for a language that makes no noise to whisper what is neither living nor dead. All words are too loud, too rapid, too sure, I'm searching for the names of the shadows *between the words*. (Cixous & Calle-Gruber, 1997, p. 152)

The notion of shadow is not limited by its metaphorical link to the interplay between light and dark, as it has mostly been employed thus far in this text. One also speaks of the shadow in dream analysis—the repressed and unacceptable tendencies, or the dark side, of our nature that have not yet been integrated into our egos.

only the shadow knows...

as Jung tells us
the shadow
is the part that we deny
toss over our shoulders
keep it down on the ground where we think
it belongs

the shadow

is seldom pondered unless one is cast onto your page as you scold
'don't read over my shoulder!'

or, science students are diligently constructing their solar eclipse viewers
so they can react to
'protect your eyes from the shadow!'

or, Bill Nye is disturbing the silence of your living room as his voice bellows
'have you ever noticed that your shadow is shorter at lunchtime
and longer at dinnertime?'

only the shadow knows...
but who knows (my) shadow?

the shadow

reveals so much about us...
casting a true light, so to speak,
on our dark side

but if our dark side
is that which does not obey the rules?
does not fit societal norms?
then why don't we call it
our light side?

I don't know about you
but I want to side with my dark side
for it is closer to knowing
why we are who we are
why we do what we do
why we know when we know

My shadow is tacit knowing
and I fail to know it well...
for as I turn to look
to catch it just being
my umbra shifts
and dark becomes light
light becomes dark

it still reflects my silhouette
for all to see
but me

unless I stop to reflect
on what my shadow knows

> **"All people dream..."**
> (Kelsey, 1991, p. 188)

According to O'Connor (1986), the "rejected and unacceptable tendencies cluster together to form figures, shadow figures, because they are shadows formed by the light of the other acceptable conscious qualities within our personalities" (p. 107). The shadow figures live in a shadowy, unconscious place, just barely outside the ego's field of view.

> By simply being out of sight it does not follow that the shadow is out of mind, and indeed the shadow – or more accurately, the shadow figures – go on exerting a powerful influence over our everyday life, often with us being totally unaware of this influence. (O'Connor, 1986, p. 107)

I am drawing on this particular context for shadows because I wonder if maybe 'shadow figures' are what Cixous is looking for when she says:

"I'm searching for the names of the shadows *between the words*."

I, too, am searching for ways to write about the shadows between the words (and actions) in the school mathematics and science experiences of my participants. While the collage of "Engendering Gender" speaks in the voices of my participants, the words they use are often not those directly spoken by a teacher; instead, the words are more of a ***shadow subtext*** of their experiences. My participants, like myself, experienced the **shadows** of gendered words and actions, more often than the actual words and actions themselves. These shadow figures dance about, making it problematic to understand their all-encompassing effects. They are often more **symbolic**, than material. They frequently exist in thoughts that are translated into tacit **expectations**. They cannot be addressed by merely doing and saying things differently because they exist *between* actions and words. To acknowledge the shadow existence is to acknowledge that one must **think** differently, often in uncomfortable spaces. If the shadows dance and dodge, are rejected (or ignored) and unaccepted, and mostly reside in the unsaid and materially invisible, how do we name them?

I would like to present parts of my second conversation with Elsie, to help highlight these uncomfortable spaces. To set the context, it is important to know that Elsie and I spoke in our first conversation about a particular female math teacher that she highly respected throughout high school. During the course of our first conversation, Elsie started to realize that, although the teacher made the mathematics very clear and interesting, she paid closer attention to the boys and their successes in the classroom, seldom providing any encouragement to Elsie. This is reflected in the previous collage quote by Elsie:

I don't know if this has anything to do with your research
but I loved our female math teacher.
But I always felt like she was teaching to the male students in the class.
Her favourites were always the males.
And a lot of other people noticed that too.
(Elsie, October 1999)

In our second conversation, I asked Elsie about this realization of her teacher's gendered approaches and how/if she had been thinking about it between our conversations.

conversing with Elsie...

Kathy: Did you find that when you left our last conversation... that you thought about the things we talked about...?

Elsie: Yeah, I thought about the things we had talked about because... it was *weird*, I had never really thought about them before— until we talked about them— and then after I thought about them more.

Even in lessons that I'm teaching now, I'm starting to think about *weird* things and I don't know if I *should* think about them or if I should just ignore them. I was doing a lesson and I had a list of questions at the end, just to see if they understood what I taught them or not. I asked the question and about four of them put up their hands and I started to think, "ok, I asked a girl last time, ask a boy this time." I don't know if I *should* do that or if I should just ask whoever...

Kathy: In listening to the tape of our last conversation, I was struck by one thing in particular. In the very beginning there was a particular teacher that you spoke a lot about—it was very clear that she was well-respected... she taught you math. She was one of your favourite teachers because everything was really clear...

(At this point, Elsie starts to smile and laugh, showing a kind of knowing about where I am going with this.)

... and it was interesting that, from my perspective in listening to the tape, that as we talked about her more, I started to get the idea that you were looking more closely at some of the things she did... started kind of questioning things.

Elsie: Exactly, yeah. It was weird... she was somebody I always looked up to, and not that I don't now but... maybe differently... I'd love to get in touch with her now and let her see where my life has ended up.

And like I said, in the classes now that I'm involved in, I don't know if I'm thinking *too much* or trying to be *too aware* on that level. I hope not. It's just funny that I never made that connection until now... and if I had never been part of these conversations with you, would I *ever* have seen that connection?

Kathy: How does that connection help you, now in the present or in the future as a teacher?

Elsie: I think as a teacher... being more aware of my students and their needs and their interests... I think *everybody*, not just specifically trying to impress the boys... because I *could* possibly end up doing some of the same things she did... if I put myself in her situation, there would be similarities I think.

I think Elsie finds herself in the uncomfortable spaces between what is said and what is unsaid; between becoming critically aware of the relationship between said/unsaid and a fear of thinking too much and becoming too aware. She clearly expresses discomfort, as she speaks about questioning everything she does, wondering if she is becoming caught in the tension between what she says and what she does not say. While she is uncomfortable with having noticed such a space, the very noticing has caused her to shift and re-position herself such that it is unlikely she could ever return to her previous position of 'unawareness' and 'non-questioning'. She is acknowledging the shadows between the words and actions that, in many cases, speak louder than the words and action themselves. The spaces between— where shadow is formed by the interplay of light and dark— is a space of doubling, of shifting with/in a tension between said and unsaid.

The objects of textbooks, teachers, and gendered ideology, discussed in this chapter, are not opaque. More light could shine through, between the cracks so to speak, if one is willing to critically question some underlying ideologies. First of all, as discussed in the gendered ideology section, if different expectations for males and females are perpetuated (whether intended or not), shadows will be cast. The underlying message articulates how particular people (or, gender?) are just naturally more able to do mathematics and science than others. Often this division of *good at it/not good at it* is focused along gender lines but it also works to create pervasive epistemological divisions based on shadows of what it means to know. It follows then that another aspect of critical questioning of ideologies must be with respect to *what it means to know.* Curriculum and associated textbooks often establish, unequivocally, what it means to know and how to measure it.

In teacher education programs, it is critical to examine the embedded structures that give rise to shadow casting in the classroom. This can be achieved through a careful exploration of preservice teachers' experiences of learning (in) mathematics and science and an acknowledgement of where feelings of anxiety and lack of self-confidence in teaching mathematics and science originate. If they originate in the shadows between the words and actions in the classroom, then the critical questioning must necessarily lead to an (un)learning and (un)knowing— to a widening of the cracks, allowing more light to enter and reveal the problematic nature of learning (in) math and science. As Dorey (1949) asks:

> Without the dark how understand the light?
> Without the shadow how could life be bright?
> (p. 2)

It may be that the deeper we look for the origins of attitudes and beliefs about mathematics and science teaching and learning, the less we will want to see. It is undeniably more comfortable to focus on content knowledge deficiencies and feelings of poor self-efficacy of elementary preservice teachers then it is to imagine that current classroom experiences and idea(l)s for learning mathematics and science are grounded in a very shadowy texture of unquestioned ideologies. The exclamation of "oh, the boys will love this"— made by the elementary science teacher written about in the introduction to this book—cuts directly to the core of gendered ideology in its nescience of the **shadows** between the words, between the actions, between the thoughts.

How Does Light Travel?"
Review Word Search with Hidden Message

N	O	I	T	C	A	R	F	E	R	T	H	G	I	L	T	T
T	T	I	N	L	A	N	G	U	A	G	E	M	E	E	N	E
A	R	E	C	T	I	L	I	N	E	A	R	T	X	O	E	X
C	R	E	I	D	E	O	L	O	G	Y	F	T	L	E	R	T
I	C	T	H	Y	O	R	E	L	C	I	T	R	A	P	A	B
T	N	Y	D	O	O	U	F	E	C	L	I	P	S	E	P	O
T	R	A	N	S	L	U	C	E	N	T	R	T	E	P	S	O
N	H	N	A	C	T	O	H	I	R	N	G	A	N	R	N	K
S	O	E	O	D	L	L	G	E	A	E	R	N	I	O	A	F
M	S	I	V	I	T	C	U	R	T	S	N	O	C	P	R	L
N	G	S	T	A	T	I	N	A	A	R	M	C	K	A	T	A
P	H	Q	C	P	W	C	F	O	C	P	N	X	E	G	U	T
N	L	S	E	I	R	O	E	H	T	U	H	Q	U	A	M	L
W	O	D	A	H	S	O	I	L	C	O	O	Y	Q	T	B	A
G	N	I	W	O	N	K	S	E	F	E	H	F	A	I	R	N
A	R	B	M	U	N	E	P	B	F	E	R	P	P	O	A	D
H	E	G	E	M	O	N	Y	K	A	X	R	T	O	N	W	J

ABSORPTION	LANGUAGE	SHADOW
CONSTRUCTIVISM	LIGHT	SHADY
ECLIPSE	OPAQUE	TACIT
FLATLAND	PARTICLE	TEXT
FOUCAULT	PENUMBRA	TEXTBOOK
HEGEMONY	PHOTON	THEORIES
HOLOGRAPHY	PROPAGATION	TRANSLUCENT
~~IDEOLOGY~~	RECTILINEAR	TRANSPARENT
INTERFERENCE	REFLECTION	UMBRA
KNOWING	REFRACTION	WAVE

Hidden Message: _ _ _ _   _ _   _ _ _ _ _ _ _ _ _   _ _

_ _ _ _   _ _ _ _ _ _ _ _ _ _   _ _ _   _ _ _ _ _ _ _ _ _ _

What happens when light hits different surfaces?

Upon this gifted age, in its dark hour,
Rains from the sky a meteoric shower
Of facts... they lie unquestioned,
uncombined.
Wisdom enough to leech us of our ill
Is daily spun; but there exists no loom
To weave it into fabric...
(Millay, 1941, p. 140)

WHAT HAPPENS WHEN LIGHT HITS DIFFERENT SURFACES?

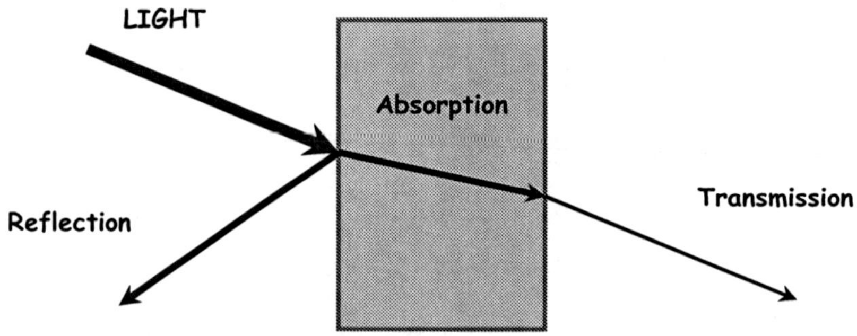

preVIEW...

In this chapter, the metaphor of light is used to discuss what happens when ideas about knowing (in) mathematics and science are discussed with elementary preservice teachers. The metaphor of light (as illustrated in the diagram above) highlights similarities between what happens when light is incident upon different surfaces, noticing how this has little to do with the properties of light itself and more to do with the nature of the 'surfaces' with which light interacts. Thus, the intent of this chapter is not so much to focus on the details of preservice teachers' mathematics and science knowledge as it is to understand how they communicate about/on this knowledge. To achieve this, the chapter elaborates on the metaphorical properties of reflection, absorption, and transmission & refraction of light.

Harriet, Evelyn, and Elsie converse about knowing...

As part of a focus group session, the research participants were arranged into small groups and asked to read a science scenario about lunar eclipses. They were given a few questions to discuss together, before coming back to the larger group of eight people.

The scenario (also presented in the newsletter in chapter two) is:
A person from your community (someone you don't know very well) discovers that you are a grade four teacher. S(he) is confused about something in science and says "oh, I should ask you… *you'll* know the answer." S(he) is trying to figure out what the lunar eclipse last week was all about—what causes it and why doesn't it happen very often. S(he) comes to you with the question.

The small group conversation that follows is between Harriet, Evelyn, and Elsie.

Discussion Question: What is your reaction to this scenario?

Harriet: My first reaction would be … what I would be thinking is 'oh, crap!'

Evelyn: Would that be your *first* reaction?

Elsie: My first reaction would be 'oh, jeez' I wouldn't know what I was talking about… well, I still wouldn't know what I was talking about

Evelyn: I'd have to get paper and start drawing things out—'here's the sun, and here's the moon'—and then probably about midway I'd realize that I don't know *exactly* what I'm talking about… but hopefully she doesn't know either. (laugh)

Harriet: And if you say it with enough conviction, people will believe you. (laugh)

Elsie: So you would attempt it?

Evelyn: Oh, yeah. I would start explaining it. I'd probably just give the basic definition of the sun's shadow on the moon and hope she didn't want any more information.

Harriet: That's why I would think 'oh, crap!' I can give you the basic shadow definition but anything more than that, I'd be like, 'well, you know what, I don't know.'

Elsie: I'd hope that maybe between the two of us, we could figure it out.

Evelyn: I'd probably just see if I had any resources with me…'oh, I've got it in the book here, look.' (laugh)

Harriet: The handy, dandy textbook… just have a big bag full of everything… all the curriculum… 'let me confirm and get back to you.'

Discussion Question: Why do you think you react this way?

Harriet: For me, I know it's because I don't want to be wrong, because I'm a teacher. I think it's because I don't want the woman going back to the community or someone else and going '*well*, she's a grade four teacher, and she can't even explain the lunar eclipse… what a moron.' As a teacher…

Elsie: [we're] supposed to be all knowing

Harriet: Yeah, yeah.

Elsie: I always second-guess myself. Even if I think I know something, I'll start explaining it and then I'll go 'no that's not right.' Even though it could be identical... you know?

Harriet: I do that too.

Evelyn: I would think sort of the same thing. I would think that because I'm a teacher I should *know* the answer.

Harriet: I think that's a general message... the teacher knows... I think that's so stressful.

Discussion Question: What do you think is her/his perception of you before s(he) asks the question? ... after s(he) asks the question?

Harriet: I think for me the perception would be 'yeah, you know, you're great' and then after I explain it, she might be... again, it goes back to 'you're such a moron'. Like, 'you can't even explain a simple definition, how can you teach grade four?!' I think... I don't know, it would be good before and then after... I don't know.

Elsie: I guess it would have to be good before if she was going to ask you.

Harriet: Yeah, she probably would think you're fairly smart.

Evelyn: You know, you don't ask questions to somebody if you think they're going to give you the wrong answer.

Harriet: And *then* you give them the answer and they think 'oh good! You got your degree from *which* Cracker Jack box?'

Elsie: Well, maybe you'd just realize that you have more in common... like, *both* of you don't know.

Harriet: See, for me, if I couldn't explain it, I'd totally think she'd think I was such a moron. I mean, if you [Elsie] think 'in common' then that's [good], but for me... I'd probably be apologizing because I didn't know. I'd be like 'I'm just so sorry.' I know I would... I'd be excusing my explanation.

Evelyn: You don't "*do*" lunar eclipses in grade four. (laugh)

Harriet: 'Sorry, that's not in the curriculum.' (laugh)

Harriet: But if you [Elsie] think you'd have more in common...

Elsie: Well, not so much in common maybe but... 'oh, you don't know either?'

Evelyn: See, I would get the right answer no matter what. Even if I was researching for the next *three* days, I would get the right answer...

Elsie: ...go and find out and then...

Evelyn: She'd be getting pages and pages on lunar eclipses; she'd never want to ask me again.

Elsie: (laugh) yeah… she would eventually…

Evelyn: She would eventually find out the *answer*… but a lot more than she'd ever want to know.

Harriet: 'I'm never going to ask that teacher again!' (laugh)

Elsie: …or, 'boy, is *she* ever resourceful!' (laugh)

Harriet: She'd be like, '*never mind*.' Or, we'd be like, 'Sorry. It's a secret… we can't tell you.' (laugh)

Discussion Question: What does this mean to you in terms of your 'science knowledge'?

Evelyn: I saw the lunar eclipse. I thought 'oh, that's really nice' and then I went into my house and went to bed. I didn't stop and think how does this happen.

Elsie: I had to go to the observatory and look at it and ask all the people who were there tons and tons of questions. And then I walked away and I was *all* confused. 'What did she say again?' But, I don't know what that means to me (laugh). I guess I have to find the answer… I don't know.

Harriet: I think, maybe, you want to *really* know the answer, you try to find it, but then I'm always confused. I don't know if I overload myself with trying to know it all or whatever. I try to learn, but then I'm confused and I'm like 'oh, well.'

Evelyn: I guess for me, then, my science knowledge only comes into play when somebody else needs it.

Harriet: Yeah. I don't go around thinking…

Evelyn: As long as it works for me, I'm ok.

In reading through this conversation, I am struck by the realization that my participant preservice teachers react and respond very differently when facing questions that 'test' their knowledge. At one point, Elsie stated that she would just respond with 'I don't know', hoping that it might be possible to work out the answer together with the other person. Harriet expressed that such an approach would not work for her, since she felt very strong pressure to know the answer, as a teacher with a certain public image. Both Harriet and Evelyn expressed the need to either dodge the question completely or to 'buy some time' while they secretly tried to absorb all the knowledge they could and then return with all the answers.

In this chapter, I will endeavour to discuss how my participants perceive, and communicate about, their knowledge of/in mathematics and science. The chart below summarizes how the metaphor of light (incident on a surface) corresponds with some of the ways in which my participants communicate about knowing in mathematics and science. I will elaborate on each of these comparisons in this chapter.

Light (metaphorically speaking)	**Perceptions Of Knowing**	**Belenky et al.** (1986) (*Womens' Ways of Knowing*)	'Other' Theories
Reflection	Resistance; 'don't know' 'not good at'	Silence	Michael (1996) ("discourses of ignorance")
Absorption	Memorize rules; regurgitate definitions	Received knowing	
Transmission & Refraction	Transform knowledge; bend it; make it personal	Constructed knowing	Polanyi (1966) ("triadic" knowing)

For various reasons, I am reluctant to cite the above resources and to introduce my own categories as displayed in the 'chart'. When one introduces categories, there is often an assumption that the author claims to have dis/uncovered a universal principle— to have calmed the chaos. (At least, in *my* critical reading, this is my usual criticism of categories). I cite Belenky et al. (1986) and Michael (1996) because their theories encourage me to reflect on the connections between math/science and knowing; to reflect on theories of knowing and theories of knowing math/science so that I might better understand how spaces could be created for personal knowledge in the knowing of math/science. It is important to emphasize here the relevance of the metaphor of light being incident on different surfaces because light is seldom completely reflected, transmitted, or absorbed. In fact, each of these phenomena occurs simultaneously and partially, thereby giving different readings on light intensity at different points on the path of travel. Therefore, one might best imagine a blurring of the categories, both for light and for knowing.

Word Play...

light meter n. an instrument for measuring the intensity of ***the light***, esp. to show the correct ***photographic*** exposure

knowledge meter n. an instrument for measuring the intensity of ***knowledge***, esp. to show the correct ***educational*** exposure

intensity n. the measurable amount of some quality, e.g. brightness

exposure n. a physical condition resulting from being exposed to the elements

Before delving into this metaphorical arrangement and the corresponding theories outlined in the chart, I would like to expand on my thinking about the use of this light metaphor in the context of this chapter. I would certainly be falling into a trap of objectifying learners and encouraging enlightenment notions if I adhere too closely to the metaphor of 'light incident on a surface'. I do believe, however, that the terms **absorption**, **reflection**, and **transmission** & **refraction** represent phenomena that call to mind, metaphorically, what happens (or could happen) in learning mathematics and science and in *communicating about* knowing mathematics and science. As I address each of these optics terms in separate sections of this chapter, I hope to clarify the more-than-literal meanings I am attributing to each of these words. I would also like to clarify my intended (mis)uses of the terminology associated with the theories outlined in the chart. I am acquiring the parts of the theories of Belenky et al. (1986) and Michael (1996) which I believe inform me in the context of my research. I am deliberately choosing not to acquire (and perhaps even be critical of) other parts of the theories which I believe have over-simplified and generalized far beyond any goals I would have as a researcher attempting to 'classify' and 'understand' knowers.

in presence
there is
absence

ABSORPTION

I would like to first discuss the notions of absorption and absorb…

ab·sorp·tion (eb-sôrp[1]shen, -zôrp[1]-) *noun*

1. The act or process of absorbing or the condition of being absorbed.
2. A state of mental concentration.

ab·sorb (eb-sôrb[1], -zôrb[1]) *verb, transitive*

ab·sorbed, **ab·sorb·ing**, **ab·sorbs**

1. To take (something) in through or as through pores or interstices.
2. To occupy the full attention, interest, or time of; engross. See synonyms at monopolize.
3. *Physics.* To retain (radiation or sound, for example) wholly, without reflection or transmission.
4. To take in; assimilate: *immigrants who were absorbed into the social mainstream.*
5. To receive (an impulse) without echo or recoil: *a fabric that absorbs sound; a bumper that absorbs impact.*

(AHDEL, 1992)

since I believe these words connect closely with ideas discussed in the previous chapter on the rectilinear propagation of light. Several of my participants referred to their experiences in school mathematics and science as mostly rote learning without enough (or any) personal meaning or understanding for them. As with most rote learning experiences, the degree of absorption of the material is usually verified through the regurgitation of the facts on a written test. An important question is, of course, whether any learning actually occurs when the educational process has consisted mainly of absorbing material and then returning it to the sender unchanged. A pleased student often boasts of her learning: "I know those definitions *by heart*." But I find myself pondering whether the heart really has anything to do with it.

I have the memory capacity to memorize the answers
to the questions in the textbook
so if the same one shows up on the test, I can sometimes get it
'cause I remember the question.

I don't get it because I know how to do it.
I get it because I copied it off the board and I memorized that...
I can't do it if they shake up the test or put an application on it.

They gave you a handout.
You'd memorize it, spit it out on the test, and forget it the next day.
I'm very good at definitions.
If I *had* to have a science test, I'd want it to be that;
I can rattle it off for you.

If you look at Bloom's taxonomy... all the stuff I learned in science...
it's all right at the bottom
just knowledge level
no application

(Ursula, January 2000)

Belenky et al. (1986) define 'received knowers' as those who "feel confident about their ability to absorb and to store the truths received from others" (pp. 42-43). Received knowers "equate receiving, retaining, and returning the words of authorities with learning— at least with the kind of learning they associate with school" (p. 39). In describing a woman who has been 'classified' as a received knower, Belenky et al. state:

> She "learns" the material; that is, she stores a copy of it, first in her notes and then in her head. She does not transform the material; she files it "as is". She willingly reproduces the material on demand, as on an exam; but she feels betrayed if the teacher asks her to "apply" it or to produce materials on her own. (p. 42)

If my research has any emancipatory intent, it would be in the hope of engaging my participants in critical conversations about, and reflection on, their potential positions as received knowers. As long as a knower is unaware of her/his position as a received knower and/or unaware of any other way of 'learning' and knowing, then movement to a position of constructed knower is unlikely to occur.

To help facilitate a reflection on the active role of the knower in the acts of knowing, I conversed with my participants about their experiences of knowledge 'absorption'. It became clear to me, early in my conversations, that most of my participants were fully aware of their status as 'received knowers'. (We did not use this terminology, of course, but we spoke at great length about our experiences of copy, colour, recall, as described in the previous chapter.) As illustrated in the pilot study quotation above, the preservice teacher resents that her participation in 'knowing' was limited to receiving and memorizing. I believe that an awareness of the nature of these kinds of experiences is why several of these women preservice teachers volunteered to be a part of my study in the first place. From the early onset of my research conversations, it was made clear to me that these women were mindful of, and some even infuriated by, the passive role they played in their mathematics and science 'learning' experiences.

> it was all memorization for me
> now that I'm realizing it doesn't have to be that way
> I really am kind of resentful that...
> people didn't know better.
>
> **teachers** didn't know better.
> **why** didn't they know better?
> why did they make me **do** that?
> why did they make it so empty and meaningless?
>
> when it could be so meaningful.
> (Preservice teacher, pilot study, 1999)

> I like to be in control of the things that I learn. I want them to be *my* things to learn and if somebody would have given me just an inch in science to learn something for myself or discover something for myself, I think it would have meant something to me. It's not that I didn't have the interest; I'd go with the interest. I thought Biology was really neat and I thought that it had potential. But when I got there, someone was just giving me boring information. You know? There's no reason why it couldn't have been fun for me. Like, I *loved* the animals we had in that room—we had fish and we had a bird... I loved that stuff about it— the plants. But they didn't give us a chance. Like, you could be the person who *fed* the fish, if you were inclined, but...They were just a backdrop for somebody to give you these notes. *I never interacted with those fish or that bird for one second.* They were just there...probably because someone said a biology class should have this and this—and one of the things was a fish.
> (Ursula, October 1999)

I certainly do not believe that any one of my participants *chose* to have such meaningless experiences of school mathematics and science learning. Even as self-acknowledged received knowers, however, they were seldom able to envision how things might be different. Without direct constructivist experiences or experiences with the idea that the knower is an intimate part of the known, it seems they just accepted that their experiences reflected the nature of these subjects and/or their own personal (in)capabilities (this will be elaborated on in the next section on reflection).

In addition to the characteristic of 'absorbing' knowledge, received knowers are also often intolerant of ambiguity and prefer to rely on the comfortable belief that things are always right or wrong, true or false. As a received knower, it is difficult to reside in the metonymical space of the dualism because such a space of ambiguity makes things far more complex and messy.

> For those who adhere to the perspective of received knowledge, there are no graduations of the truth—no gray areas. Paradox is inconceivable because received knowers believe several contradictory ideas are never simultaneously in accordance with fact. (Belenky et al., 1986, p. 41)

Several conversation fragments emerged in my research that had me and my participants pondering the difficulties involved in moving from a perspective of received knowing to one of constructed knowing. As illustrated through the following quotes, my participants expressed a perceived impossibility to question science, a lack of 'understanding' of science concepts (such as chemical bonding) and a repeatedly reinforced 'norm' of definition learning and finding answers to prove one has knowledge.

I don't think we feel *capable* of questioning it [science]. I don't think it would ever cross my mind to question it because I'd feel that I don't know enough to question it. You have to *doubt* something first and if you're not given any room to doubt anything then how are you going to question it? (Ursula, October 1999)

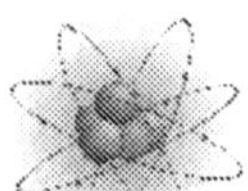

Chemistry...
I never understood.
I could memorize formulas and figure which things had how many bonds.
I could do *that* but I didn't know *why*. I still don't understand the purpose *of* bonds.
I could tell you how many they *have* but...
(Evelyn, October, 1999)

You have to have something to prove to others that you are teaching your students something. I think that's the pressure to have an answer or find an answer. It's an uncomfortable feeling if there isn't an answer or you don't know it or you can't find it. That concept of 'if you don't have an answer then you don't have the knowledge'... yeah, they go hand-in-hand. (Maude, April 2000)

It appears that my participants had very few experiences which demonstrated to them that they had an active role in the construction of knowledge. For me, their voices further emphasized the cycle of miseducation and contradiction, where mathematics and science curricula call for active participation of the learner but the experiences of learners directly contradict such calls. Why does such a schism exist between curriculum and experience, between theory and practice?

Reflexions

I respect the work done by Belenky et al. (1986) and I find their categories useful in understanding more about the experiences of knowing that my participants relate to me. Underlying all their categories and conversations about knowing, however, is a systemic fault line that somehow goes unnoticed and unquestioned, despite constant tremors beneath. I often ask myself why many researchers insist on focussing on the knowers and ***how*** *they know in order to categorize them, without questioning how anyone might become a knowing 'member' in such a category in the first place. Is it a characteristic of the knower that results in her believing UNequiVOCALly in absolute truths and authorities, or is it something about how the known is 'presented' to us, void of spaces for interacting with it? Sara Ruddick (1996) critiques Women's Ways of Knowing (WWK), saying: "Although WWK speaks of 'ways of knowing', it classifies people. A woman doesn't know in a receptive way in respect to some subjects in certain circumstances. Rather she is a 'received knower'. To the extent that her way of knowing is lesser, so is she" (p. 253).*

Ruddick counters the WWK view by remarking that "a person does not 'know' irrespective of the object of knowledge or the practices in which she is engaged" (p. 254). I wish to be very careful not to make it appear that I, too, am classifying my participants. I simply wish to explore how their experiences of knowing and how they choose to talk about these experiences demonstrate aspects of silent, received, and/or constructed ways of knowing. I especially want to clarify that my conversations can really only suggest to me certain ways of knowing with/in the specific contexts of each of these women learning mathematics and science. My participants have clearly shown, in conversing on other subject areas, that they cannot be classified as particular 'knowers' in general. Perhaps I am feeling slightly uncomfortable with a common research goal of trying to wrap everything up in a neat and tidy package. Perhaps I am just providing a caveat to the reader: Be careful how you interpret (and interpolate) my interpretations!

REFLECTION

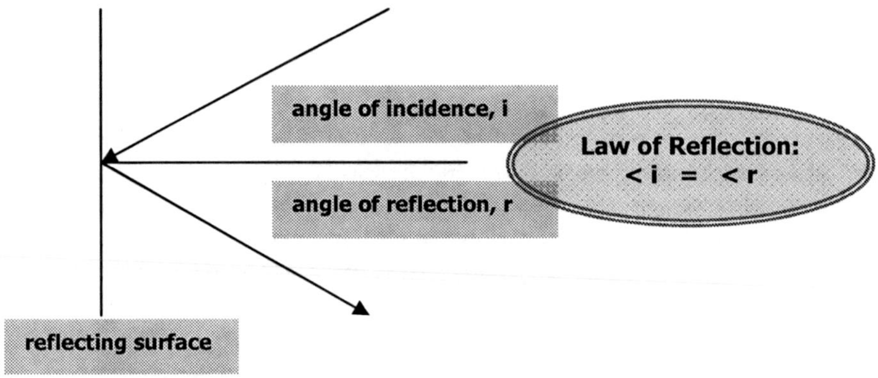

In contrast to the previous section on received knowing, this section explores the ways in which people formulate a relationship of NON-knowing with/in mathematics and science. I use the word 'reflection' as a title for this section of the text to call to mind the nature of responses which reflect the question back to the questioner by claiming to 'know nothing' or to be 'not good at' the area in question.

> repulsion, repellence
> repulsive force, centrifugal force
> repellent quality, repulsiveness, ugliness
> **reflection**, recoil
> driving off, beating off, defense
> repulse, turnoff, rebuff, snub, refusal, the cold shoulder, rejection
> brush-off, dismissal, ejection
>
> (ORTEWP, 1994)

Goldberger (1996a) defines the knowledge perspective of silence as "a position of not knowing in which the person feels voiceless, powerless, and mindless" (p. 4). This knowledge perspective has been predominantly explored in Belenky et al. (1986) in terms of women who view themselves as "deaf and dumb" with little or no ability to think. In other words, the silent knowers identified in *Women's Ways of Knowing* are not choosing silence but silence is somehow imposed on them through gendered, classist, or racist constraints. Belenky (1996) presents a somewhat reconstructed version of the definition of a silent knower in saying:

> The Silenced see words more as weapons than as a means of passing meaning back and forth between people. They do not believe themselves capable of understanding and remembering what the authorities or anyone else might say to them; they do not feel capable of articulating their own thoughts and feelings to others. ... Without the most basic tools for dialogue the Silenced feel voiceless and excluded from the social life of their community. (p. 394)

As suggested through the re-presentation of the voices of my participants, I believe that silence is a much more complicated phenomenon than described in

Women's Ways of Knowing and also as above in Belenky (1996). The complication enters this neatly defined category when the question is asked whether silence as a knowledge perspective is imposed or chosen. Even as I write this, having pondered the question for some time now and having directly asked my participants about their tendency toward 'silence' when talking about mathematics and science knowing, I am just beginning to embrace the complexity of the question. I am once again tossed into the metonymical spaces of doubling, where knowers **neither** choose silence **nor** is it imposed upon them but it is **both** chosen **and** imposed in a very messy intertwined manner.

Michael (1996) conducted a study in which he asked people (public, lay persons) to reflect upon the epistemological status of their scientific knowledge. His study is a welcome change (for me) from the alternative conceptions in science studies discussed earlier in this text. Traditional science education research on alternative conceptions addresses a need for *changing* misconceptions so that both students' and teachers' ideas about science will be more in line with 'official' scientific knowledge. The research literature often addresses this need by extracting information from participants on their knowledge (a mechanistic and instrumental approach), without considering the value inherent in asking them to engage in reflection upon their own knowledge. In our western culture, which highly values correct and official knowledge, such a form of exploration (exploitation?) into alternative conceptions can be a very intimidating exercise. This is especially true for preservice teachers who (think they) have much to lose in the way of respect in our school culture if they are to demonstrate anything *but* official knowledge. Take, for example, Sylvia's response to being asked how she would feel about not being able to 'get the answer' in the back of the book when doing a math problem in front of the class:

> If they were under grade 5, I wouldn't be worried. I don't think they would think any *less* of us. But I think if you start going into middle years and high school, and you came up with the wrong answer, they'd be like
>
> ***'yeah, you don't know anything'***
>
> I'm nervous about middle years so maybe that's why I'm saying that. I don't know. I'd be more nervous for sure. And that's one reason why I'm not a high school—*especially* a high school teacher—because I'm nervous about teaching math and science. I won't even *go* there because I think they'd challenge me so much that I'd be constantly checking myself. I don't think I could do it.
>
> (Sylvia, January 2000)

In asking people to reflect on the scientific knowledge they possess, Michael (1996) found that "they do not simply have a 'defective' body of quasi-scientific knowledge, [but that] they have none at all" (pp. 107-108). In other words, the people interviewed in Michael's study reflected upon their *absence* of scientific knowledge such that Michael identifies "discourses of ignorance that people mobilise when reflexively commenting upon their lack of scientific knowledge" (p. 108).

Inside Research...

> Discourses of ignorance not only constitute a means of understanding and explaining one's lack of knowledge, they also signify and reflect the perceived social relations between science and lay person, between self and expert. As such, they are connected with questions of social or cultural identity, and any educational programme must address what the knowledge it disseminates implies for these socially embedded conceptions of self and other. (Michael, 1996, p. 112)

(Do educational programmes even acknowledge, let alone address, the implications of knowledge on self?)

One key aspect of Michael's work for my research is in his statement that "ignorance cannot be treated as simple deficit; it entails active construction" (p. 111). In other words, people actively formulate a relationship with/to science.

> On the substantive, theoretical level, we can imagine how lay people's self-ascriptions (of ignorance) set up a relationship between themselves and science—one in which the causes or reasons behind their 'ignorance' reflect their social identity and their relations of dependence, co-operation, or challenge—that is relations of power vis-à-vis science. (p. 112)

Michael discusses three discourses of ignorance in which absence of knowledge in science (in particular, in his study, about the science topic of radiation) is actively constructed:

1. 'Ignorance' and mental constitution. 'Not scientifically minded'; relation of *dependence on* science
2. 'Ignorance' and the division of labour. 'Not my job'; relation of *co-operation with* science
3. 'Ignorance' as deliberate choice. 'Not interested/relevant'; relation of *challenge to* science

All three of Michael's (1996) discourses are insightful ways to view the construction of a person's relation to science. In many respects, all three manifest themselves to varying degrees in my research conversations. Glimpses of the second discourse— 'ignorance' and the division of labour— have been detected occasionally in my conversations. For instance, one of my participants dismissed a particular science concept as not being important for her to know because the concept 'belongs' in the grade five curriculum and she plans to teach grade two, indicating that she felt it was not her 'job' to know and teach it. The third discourse— 'ignorance' as deliberate choice— was most evident when my participants spoke of their *classmates*, who expressed an unambiguous dislike of science and were adamant about intentions to not teach it if they possibly have a choice in the matter. (These preservice teachers—the classmates— did not volunteer to converse with me about their mathematics and science experiences but it would certainly be an interesting follow-up study). In some respects, I think Evelyn was using the discourse of 'ignorance as deliberate choice' when she expressed her lack of interest in knowing (about) science, unless "somebody else needs it". In her account, she watched the lunar eclipse and then went to bed, without really caring how or why the eclipse happened.

I believe that the second and third discourses are not as pervasive in my study since the preservice teachers who volunteered to converse with me did not *usually* declare that knowing science is not their job or that they choose not to be interested in it because of its irrelevance. I also believe that, in the case of teacher education programs, the three discourses are not so separable, as Michael (1996) indicates in his research. In fact, I think the second and third discourses represent epistemological perspectives that are often special cases of the first discourse. For instance, Harriet shared with Evelyn and Elsie (in the conversation beginning this chapter) that one way she would mask her lack of knowledge about eclipses might be to explain that the topic is not part of her grade four curriculum. In other words, the fact that she might ask lots of questions and really want to know the answer might be irrelevant in the final analysis, since she may still end up explaining her lack of answers in terms of her job requirements. Who really knows how she feels about her 'mental constitution' when she is simply trying to *save face*?

In this research, I am especially interested in focussing on the first discourse: 'Ignorance' and mental constitution. I am particularly intrigued by this discourse because I believe it is intimately connected to the math/science (and) knowing relationships formed by my participants. During my research observations and conversations (as well as my years of tutoring mathematics and science), I found it uncanny the number of times my questions relating to knowing in mathematics and science were reflected back to me with a statement about not knowing mathematics and science because of mental constitution, or (in)capabilities.

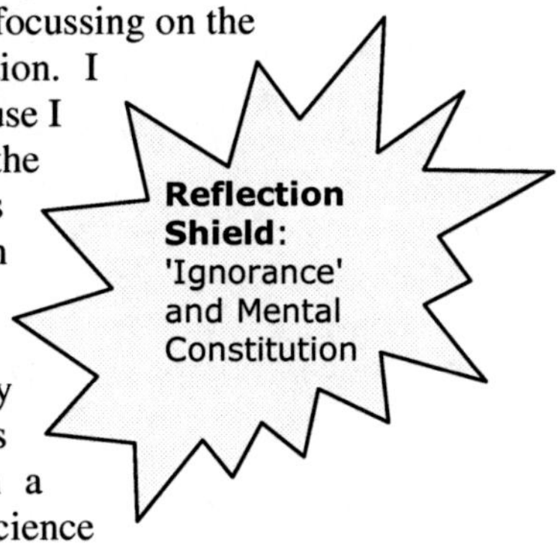

I don't know a thing about it.
I can honestly say I know nothing.
I know nothing about science.
Like I said, if I didn't have to take science I wouldn't.

(Preservice teacher, pilot study, April 1999)

If I don't get something in math and science, I'm stupid.
Like it's not 'oh, it's a hard class' or 'it's a hard idea'.
I'm stupid. That's all there is to it.
My strength is languages. If there's something I don't get,
I'm like 'well, I need a different explanation' or 'I need help with this'.
But with science— 'I'm just stupid. I'm going to give up.'
(Helen, November 1999)

With something like English, I know
I'm strong enough to *get* the answer.
But science...
I just have no clue.
So I'm less likely to risk being wrong.
I accept it as a fact actually (laugh) [that with science]
it is more likely I'll be wrong.

I think a lot of people don't feel that there's
room for speculation in science;
that it's just a 'yes' or a 'no'.

(Evelyn, December 1999)

During a physical science workshop in an elementary education class, I wrote a few notes while observing the elementary preservice teachers working in groups to understand more about the model of the atom. It was quite astonishing to me at the time to hear so many comments relating to their perceived 'absence' of scientific knowledge. Comments such as "I'm not very good at this" and "I don't know anything about this" and "I'm not scientifically minded" were commonly articulated as I walked around the room.

As Michael (1996) reminds me: "The relation of dependence to science is partly born of exclusion from it" (p. 117). This exclusion is usually expressed in terms of the person's own (in)capacities, announcing at the outset of any apparent 'testing' of their scientific knowledge "that they are constitutionally not mentally equipped to fathom the mysteries of science" (p. 117).

Persons adopting this discourse about 'ignorance' and mental constitution will also attempt to explain their exclusion from science and quasi-dependence on it in terms of the nature of science. One person interviewed in Michael's (1996) study defined the subject areas that she had the least amount of insight into as science; for her, science became self-defined as those areas that were important but also seemingly unrelated to the nature of her everyday life. In my research, my participants pondered the idea of everyday science, recognizing that they envisioned a schism between 'real' (school) science and the science of their everyday lives. The following conversation excerpt, between Sylvia, Harriet, and me highlights this:

Conversing with Sylvia and Harriet

K: I know what you mean, science can sometimes appear to be way out there, and you don't feel a part of it…
H: It's like this little sub-culture and I'm on the outside… at the fence; you know, I can see it but I don't know how to get to it. It's way over yonder and you think 'if I could just get over the fence and start walking to it', then maybe…
K: And what about if it started walking toward you?
H: I'd have a *heart attack*.
S: Oh god.
K: Don't you think a lot of the public image is 'oh, science is not for me'?
S: Yeah, and I never even thought of that… science coming to *me*?! Yeah right, I have to always go to science. And yet, why not?
K: But science *does* come to you every day, doesn't it?
H: Maybe it's because we don't see it *as* science.
S: Maybe it's because the questions asked aren't the same questions that *I* would ask or not the same way I would look at everything. So it always seems that I have to go to science. But maybe if different questions would be asked or things were approached in different ways, maybe science would come to me more. I don't know.
H: I'd probably have a heart attack if it did. I'd be like 'whoa, something's really wrong with the world…'
S: I'd think 'ok, that's science but it's not the *right* science.' It's science but it's not the lab coat guy in glasses.
H: I would think 'oh, that's too easy'. 'It's not the *real* science.' I would think 'it's wrong'.
S: 'That can't be real science'. Yeah, 'it makes *sense*; I get that— it can't be *real* science.'

all of a sudden
i wonder
is ***knowing by heart***
like
knowing by heart attack?

in presence there is absence

Taken together, the two possible expressions of 'ignorance' (i.e., the person's perceived and mental constitution (in)capacities and the perceived complex nature of science) set up a dangerous cycle of exclusion from science: "I know nothing about it because it's science" or "it's science because I know nothing about it". In addition, exclusion is also a characteristic of the silence(d) way of knowing, as illustrated previously in Belenky's (1996) words: "Without the most basic tools for dialogue the Silenced feel voiceless and excluded from the social life of their community" (p. 394). Again, I believe it is extremely important not to overlook the context. I would be amiss in not emphasizing that I am identifying characteristics of silence(d) knowing in my participants as related to knowing in mathematics and science. Exclusion from the "social life of their community" refers, in my study, to exclusion from the mathematical and scientific communities. It is the notion of EXCLUSION…

ex·clu·sion (îk-skl¡[1]zhen) *noun*
1. The act or practice of excluding.
2. The condition or fact of being excluded.

ex·clude (îk-skl¡d[1]) *verb, transitive*
ex·clud·ed, ex·clud·ing, ex·cludes
1. To prevent from entering; keep out; bar: *a jar sealed to exclude outside air; an immigration policy that excludes undesirables.*
2. To prevent from being included, considered, or accepted; reject: *The court excluded the improperly obtained evidence.*
3. To put out; expel.

(AHDEL, 1992)

and its connection to the types of relationships my participants formulate with/to knowledge that I am particularly interested in understanding. On the surface, the theories discussed by Belenky et al (1986) and Michael (1996) appear to illuminate how and why people in general (and perhaps women specifically) create certain relationships with knowledge when they feel excluded from/by it. From the pieces of conversations that I have (re)presented thus far, I believe that several of my participants demonstrate characteristics of the silenced knower and signs that they construct 'ignorance' because of how they perceive their mental constitution.

Inside (my) Research

As a point of clarification, I must add that I really have **very little knowledge** of how my participants *actually* perceive their mental constitution; I can only try to understand how they choose to *communicate* how they perceive their mental constitution. Yikes! I think the reader has just entered into a triple hermeneutic zone: the reader is interpreting my interpretation of my participants' interpretations of their perceptions of themselves as mathematics and science knowers!

[See **'I don't know but…'** on the following page]

'I don't know but'...

Right away, when that Tangram was in front of me, I said to somebody 'I can't do these'. I didn't know that yet! I just say it because it's easier… you have a safety net, then everybody knows you can't do it and if you're the only one at the table who can't do it then it doesn't matter, you know?

(Ursula, January 2000)

I think, in general, there are a lot more perfectionists out there than in the past because people are striving for such high goals and they're getting pressure from all sides, themselves included.

And so to come across as not knowing, you better admit it up front or else someone may turn on to you and say 'well, you don't know what you're talking about' and you'd have to admit it afterwards...

Like, 'you're an imposter' or that, 'you're faking it and I caught you'.
So... 'you can't catch me, I'm admitting up front I don't know, but this is what I think.'

(Maude, December 1999)

A lot of people would say 'I don't know very much about this. I can tell you what I remember. It may not be *right* but...'

(Elsie, November 1999)

'I don't know but...'.
I always have to say,
'I'm not sure if this is right but...'
I'm always questioning myself before I even get my answers out.

(Sylvia, December 1999)

For science I say,
'ok,
this is my thought,
I don't have *any* backing on it,
but it's still my thought'

(Roberta, December 1999)

If I'm with people who I think have more knowledge than me in something—and I don't even feel I can control it because when I think about it later I say 'why did I behave that way?'—but if I think they're going to ask me a question or test my knowledge in some way, I want to qualify right away that I'm not sure. If possible, I will dodge it completely, saying "I don't know'. It's like a wall goes up as soon as I think they're testing me.

(Kathy, December 1999)

I don't believe, however, that the roles people play in silence(d) knowing or 'ignorance' as mental constitution are as passive as they may appear on first glance. The Silenced is "a characterization that more accurately reflects how some people may be driven to a defensive posture of passivity and silence out of fear and threat" (Goldberger, 1996b, p. 346).

> Silence is a powerful weapon when it can be controlled. It is akin to camouflaging oneself when at war in an open field; playing the possum at strategic times cause the power of the silent one to be underestimated. (Hurtado, 1996, p. 382)

> Our participants draw upon their social identities to resource their understanding of their own understandings, and of their relation to scientific knowledge and its institutions. ... given the ambiguous character of science—at once monolithic and non-unitary, radically 'Other' and incorporated into the modernist self, science in its various forms can begin to impact upon and potentially affect those social identities. ... what emerges is the implicit discursive formulation of the relations of power between self and science. (Michael, 1996, p. 123)

Do these quotes hint at the fact that there is often another entire layer beneath the surface of this admission of 'ignorance' or not knowing? I think so. During a presentation of some of 'the pieces' of my research at a university seminar, an audience member contributed response data in the form of the following question:

Question: When you talked about ideology, I got thinking about Foucault's work on power. Power is a lot more complex than somebody does and somebody doesn't have the power. If you don't have power, then you find interesting ways to resist. I wonder if you find examples of this in your research... that these students find interesting ways to resist?

(non)**Answer**:
I think that— even from my personal experience— I can see this...
I can see that if you feel threatened in a certain way with something that you don't know or don't want to go there then you can come up with different ways to resist.

I 'answered' the question in a very obscure manner because the truth was, I hadn't really thought about it at the time. If resistance can be understood as contradictory processes at work (Apple, 1995) then it seems necessary for me to spend a considerable amount of time and space discussing this concept in my research. At various levels, the conversations with my participants reminded me of the inevitable contradictions involved in understanding more about the nature of knowing as perceived by individual knowers. For me, it is also inevitable to see these individuals as gendered beings in a patriarchal system.

In the following few pages, I re-present several textual pieces from participant conversations; these pieces highlight the emergent nature of such contradictions as the conversations with Sylvia, Harriet, and Helen evolved over the course of the research.

Inside (my) Research

In the first conversation with each of my participants, characteristics of 'silenced' knowing and the discourses of 'ignorance' emerged as what I perceived as my participants' personal insecurities with their knowledge levels. I pondered the transcripts of these first conversations for some time. In the second conversation, I asked each of my participants about this notion of prefacing science-related statements with a precautionary "I don't know but..." or adding on the trailer "... but I don't know very much about this" to science-related statements. We both reflected on why we thought this was such a recurring phenomenon in their experiences of 'talking (about) science (knowing)'. We spoke at great length about the importance of being comfortable with knowing and trying to remain in a personal comfort zone as a teacher. We both went away and pondered this second conversation. Then, in the third conversation, we reflected on what it means to them (and to me) to live with/in these epistemological spaces as teachers.

Conversing with Sylvia and Harriet

H: Doing the atom [workshop]... I say 'I think this is wrong *but...*'. I *know* I say that.
S: I know. I do. I don't know if that's my personality or my trait as a woman. I don't know because I noticed that a lot... 'I don't know *but...*' I always have to say, 'I'm not sure if this is right *but...*' I'm always questioning myself before I even get my answers out. And personally I *believe* my answer's right but I don't think everyone else does so I always start...
H: It's almost like a safety net. If I'm wrong, well...
S: If I make a statement, I have to put that safety net there, *just in case*.
K: That idea of a safety net... what do you see it as being safe from?
H: Being said 'oh, you're such an idiot'... because I'm just saying 'oh, this might be wrong' and if it *is* wrong than I *told* them it was wrong so they can't say 'oh, you're such a moron, you thought that was right'. Then if I *am* right, they're like 'oh, way to go buddy'
S: I think it's like a protection. It's definitely more predominant in science... like, in chemistry class.
K: There is something about warning someone that you might be wrong, rather than...
S: ...going out on a limb and then having the branch break

H: Our classmates all do that too…'I might be wrong but' or 'this is just my opinion but'
K: Do you do that with each other?
S: Oh yeah… well, it's different though. I can say 'I don't know' but I know she's still valuing what I say.
K: The word you use—'valuing'—makes me think that maybe it's not even so much about being *wrong*, as it is…
H: Being valued, maybe.

Conversing with Helen

Conversation #1 with Helen…

K: Can you think about why you give yourself that description of being good at something or *not* being good at something?
H: I don't know if what I'm thinking is what you're looking for but… I think the reason *I* do it is to *explain* a lot of things. I think I'm good at languages and stuff, and I've had reinforcement of that. High School Grade 12 English, 97%; that shows me I'm good at it. But my high school Grade 11 Geo-Trig, 76%; I'm *not* good at that. Sometimes I use it to explain things, like 'I got a bad mark; I must be really bad at it' or 'I got a good mark; I guess I'm good at that'. And I guess a lot of it has to do with how comfortable I am with something. If I'm really comfortable with something, I'm probably good at it. If I'm not *as* comfortable, I'm probably bad at it. That's probably totally a bad thing to do (laugh), but I think that's why I do it.
K: The idea of classifying oneself as knowing or not knowing about subjects… putting these labels on yourself. Do you think it is, personally for you, an issue of confidence?
H: That's a good point. I don't think it's so much what I actually *know* about it—cause I think I probably know quite a lot about math and science—but I don't want to *say* that I do because I don't want to have to do it. So if I just say that I don't know anything then… 'ok, well, skip her'. Yeah, that's a good point. With language arts, ask me anything and I'll take a guess. With science… don't even ask, I don't want to do it. If I'm wrong with something in language arts, 'oh, well, oops'. But if I'm wrong with something in science, it's like 'oh, I'm stupid. I thought I didn't know, I didn't know.' It's how I feel about the subject and me doing the subject, probably not actually what I know.

Conversation #3 with Helen…

K: It was really interesting when you said that it was not so much what you know but you don't want to *say* you know because you don't want to have to do it. I *know* that feeling… it sort of cuts people off right at the very beginning. Rather than me battling my way through it and doing all this work, if I just say that I don't know then they won't bother with me anymore. Do you still feel that way?
H: Yeah, it works for a lot of things, not just science and math. Like I play the piano so if someone wanted me to play the piano in public, which is not my favourite thing to do, I'd say 'I'm not very good'… even though I'm perfectly capable of playing the piano. It's sort of like that… if you give it to me, I could do it, but it would be *hard*, I might get it *wrong*, you might think I'm *stupid* so I'm just going to say 'you should ask someone else'.
K: I was thinking of exploring that… it's a very common thing to start off by saying 'I'm not comfortable with this so I'm warning you'. It becomes kind of a

safety net. And I'm wondering if this is a form of resistance to getting into a communication about math and science... a strategy almost, to avoiding it?
H: Yeah, I thought of that in our math class... every time we do something... like today we're doing fractions. A lot of people hate fractions, so right away you hear 'oh, I can't do fractions', 'I'm terrible at fractions', 'I'm horrible at fractions', 'I'm not going to be able to do this'. Like, they don't even *try*. I do that too, just not with fractions (laugh).
K: Do you think in some ways it is a strategy... kind of stops anyone from having expectations... would that be fair to say?
H: Yeah, we were doing a word problem [in math]. I said to the rest of my group 'I'm not going to be able to do this. This is too hard. I don't do word problems.' And then, if I *couldn't* get it, my group would say 'well, ok, she doesn't do word problems' and if I *did* get it, it'd be like 'wow' (laugh). People wouldn't be *expecting* me to do it.

It appears in many respects that my participants are reifying dominant ideologies, such as the 'fact' that mathematics and science are difficult and that not everyone is expected to be able to 'do' them. Maybe they'd prefer, in some cases, not to cha(lle)nge the ideology because they can't be bothered, as with Helen. This may not mean they are unaware of the ideologies at work but perhaps they are just not interested in trying to prove something or deconstruct anything. I know from my personal perspective, the challenge of proving something is a large part of why I continued to study mathematics and science in higher education. Initially at least, I wanted to show that I could do as well as *anyone else* [read: males] in these subjects and that was a strong motivating force for me. Now I am much more critical of this approach because I believe it fits into the strategy of 'making women fit mathematics and science'.

I believe that another reason for preservice teachers to buy into the dominant ideologies is to avoid 'inevitable' failure— that is, their existing expectations of failure may only be realized in trying to do mathematics and science. And they prefer not to set themselves up for such a 'natural', expected failure— for a self-reflective 'I just knew I couldn't do it.' This then becomes a case of ideologies becoming hegemonic.

> Hegemony… refers to a process of domination by consent, in which the general population adopts a worldview that reflects the interests of the dominant classes. This analysis allows us to see how our common-sense way of looking at things, which we derive from our traditions and experiences, is constructed so that ruling practices are maintained without usually being evident to us. (Wotherspoon, 1998, p. 94)

> Ideologies become 'hegemonic', not because they change, but because circumstances change while advocates for the ideology become more and more concerned with its preservation. What causes this decline into reification and stasis is precisely the absence of reflexiveness within ideological thought, the inability to recognize its own origins and limitations, and the lack of opportunities for thinking differently. … criticality is the opposite of the hegemonic. (Burbules & Berk, 1999, p. 61)

It seems as if some of my participants created a special form of resistance— one which involves choosing not to partake in a mathematics or science communication by an up-front claim to be *not very good at it* or to *know nothing about it*. This resistance emerges in close connection with a person's identity and it could be referred to as a 'self-preservation resistance'.

self-pres·er·va·tion (sèlf´prèz´er-vâ¹shen) *noun*

1. Protection of oneself from harm or destruction.
2. The instinct for individual preservation; the innate desire to stay alive.

(AHDEL, 1992)

Thinking back to the seminar mentioned previously, and the audience member's question, I wonder again why I initially resisted paying attention the notion of resistance. I guess the notion of resistance in my research was not something I had thought about in this light. I certainly made connections between power, knowledge and ideology, but more from the perspective of the role of curriculum. I thought of the decisions about what counts as knowledge and who decides what knowledge is of most worth as being representative of the power and hegemonic issues in education. I felt that the ways in which curriculum embeds dominant ideologies excludes opportunities for teachers to actually think they could resist. Instead, I felt that teachers just accept the 'way things are' and work from within to make the best of it. In other words, it appeared on the surface to be a case of false consciousness supporting dominant ideology without realizing it or even considering the possibility that things could be different. My research intentions, emphasizing reflection and reflexivity, were focussed on 'overcoming' false consciousness. Like Patti Lather (1991), I thought of a ***resister*** as "someone so saturated with false consciousness that she could not see the 'light' being offered her" (p. 78). Now I do not *dismiss* false consciousness but I believe it is not as deeply embedded in ideological practice as I first thought. Perhaps, as a researcher, my acceptance of false consciousness would be an easy way out of examining underlying complex issues of power and knowledge at work at *all levels*.

Did you know that...?

There is quite a remarkable history of women who were openly discouraged from engaging in intellectual pursuits, especially in mathematics and the sciences. A few of these women persisted, in spite of the oppression and adverse reactions among family and colleagues. In many cases, however, the only way for women to persist was to write using a male pseudonym, in order to be taken seriously for their achievements. One such person was Sophie Germain, who published her mathematical work under the name M. LeBlanc.

This demonstrates an admirable form of resistance, and possibly the only means available at the time for trying to open doors to women. However, secretly researching mathematics and publishing with a male pseudonym seems to be working both against *and for* the existing conditions. A history of well protected hegemony.

The contradictory processes at work in this form of 'self-preservation' resistance create and shape a **dialectic relationship *between*** the underlying ideology and hegemony of reproductive processes in schooling ***and*** the implications of a resistance to the dominant ideology that is hinged on self-preservation. In other words, my participants talked to me about the feeling of exclusion from mathematics and science, the gendered oppression they felt from teachers who did not expect girls to 'naturally' enjoy or excel in mathematics and science, and the meaningless experiences they shared in learning definitions and applying rules without understanding.

I doubt anyone could deny how these kinds of experiences work to legitimize why women might reject and resist science. Giroux (1981) states: "As the dominant ideology, hegemony functions to define the meaning and limits of common-sense as well as the form and content of discourse in a society. It does so by positing certain ideas and routines as natural and universal" (p. 94). It almost seems as if these unacceptable experiences have somehow become naturalized and 'acceptable' as part of the nature of math/science learning and knowing (for women). But it need not be like this. One must develop an awareness of how to *resist the ideological hegemony* of knowing math/science, not resist knowing math/science. In a sense then, my participants may be un/knowingly buying into the dominant ideology and reproduction of school math/science as elitist, predominantly male 'domains'.

In focusing my thinking on how the ideological hegemony of mathematics and science has been 'taught' and reproduced in the schools, I had neglected the space of 'negotiation' that these women have/had in/through their school experiences. Apple (1995) states: "Whatever reproduction goes on is accomplished not only through the acceptance of hegemonic ideologies, but through opposition and resistance" (p. 23). He goes on to say that these "informal cultural resistances, this process of contestation, may act in contradictory ways that ultimately tend to be reproductive … Resistances on one level may partially reproduce the lack of control on another" (p. 23).

"If students are to understand how the dominant culture bears down on them, they will have to grasp how it reaches into the logic of their own resistance" (Giroux, 1983, p. 167). This speaks of the possibility that cultural reproduction and production could exist simultaneously through the same action— an action meant to oppose the dominant ideology but often resulting in a form of production that further reifies the perspectives of the dominant culture. An act of resistance to official ideology and its reproduction *could* represent a form of cultural production that further disables the students' positions. In other words, by resisting 'doing' and 'liking' mathematics and science, my participants may intend to show that they do not feel the need to buy into the status and dominance of these subjects. Such a choice, however, may be interpreted not as resistance but as incapability, or lack of 'mental constitution' for mathematics and science, thus unintentionally further reifying the dominant ideologies. (*Oh, what a tangled web we weave…*)

In continuing this critical discourse (and attempting to clarify my ideas outlined in the previous paragraph), I would like to introduce selective parts of my conversations with one participant, Evelyn. My first two conversations with Evelyn opened my eyes to the tensions and contradictions experienced by many women in knowing mathematics and science, particularly elementary teachers. In our first conversation, Evelyn spoke about there being a lot of weight placed on knowing science—it is a subject given high priority and status in school curriculum. Then, in our second conversation, Evelyn talked about there being a lot of support (in general) for *not* knowing science…

E: There is a lot of support for not doing good at science
If I told people I couldn't spell, they'd be horrified—

'what are you doing in education???'
If I told people I didn't understand chemistry, they'd say:
'yeah, well, you know...'
If I don't know anything about World War II, they wouldn't want me teaching, [me] not thinking it's important.
If I told them these students don't need to know about laws of gravity, most people wouldn't blink

So, in our third conversation, I asked Evelyn about this apparent contradiction and what it means to her.

K: It's a funny contradiction isn't it? There's a lot of weight placed on knowing science and yet we talked about the fact that there's also a lot of support for not knowing it. What do you think that has to do with?

Evelyn expressed the difficulty she has admitting that she doesn't know (something in) science because she feels that there is a lot of pressure to know these things as an educated person (and as a teacher especially?)…

E: I think... a lot of weight is placed on being able to know science and math. It just sort of seems like something you should know if you want to be at a certain place or [to] consider yourself an educated person, you should have these well-rounded ideas [and] areas of knowledge.

However, she said that in our society "people feed off one another"; they'll cluster in groups and support one another in not knowing and not being good at science…

E: In almost any situation you're in, if you can find somebody else in that same situation, you can feed off each other: 'we are the people over here that don't know science'. And maybe it's just that I don't know people in my life who would say 'we are the people who don't know who the Prime Minister is' and then I could say 'well, I don't know either'. But maybe it's just by chance that I know other people who don't know science.

After all, Evelyn says, science is portrayed as being difficult and it is stressed over and over again that only certain people can do it. Therefore, it is safe (even normal) to be in a group of science non-knowers…

E: But I think it goes back to the idea that only certain people can *have* scientific knowledge. I think that's stressed over and over again. I can't remember if I told you this story or not but, on the first day of my Calculus class, the teacher got up and said...
'You are the upper echelon'.

And I had personality conflicts from Geo-Trig with this teacher and I decided not to take it. So on the second day of class, I went in and had to get him to sign my form to say that I could drop it. He said...

Oh. Decided you can't handle it, hey?'

If I hadn't been angry, I probably would have stayed in the class just to *prove* he was wrong. But there was a definite idea that it was a *superior* thing to be able to do Calculus, which has no bearing in my life. And you had to take Physics, Chemistry, and Biology to **keep your doors open**, to keep all those options open. I don't know anybody who's actually really used those.

Evelyn went on to explain that she never had any intention of studying medicine and that's the only reason she felt she would need her sciences. And a lot of her friends went on to study business, and didn't need *their* sciences either. She commented that…

E: I'm not exactly sure why we needed to keep those doors open but it was definitely a status thing in my high school and in my community.

My critical feminist perspective on this is that science-knowing has been portrayed as a 'no-win' situation for her. A lot of status is definitely associated with being able to do science but, at the same time, there is a clear message that not just anyone and everyone is expected to be one of the few that can achieve this status. This is hegemony, at its finest.

E: I don't think it was a bad idea for most people to end up taking those three sciences. But the fact that there was *pressure* to do so and it was *expected* and just that it was a *status* thing wasn't a good way to look at science.

Inside (my) Research

When Evelyn and I first spoke about these contradictions (pressure to know, support for not knowing), I think I was ready to dismiss her words as representing a confused moment. I thought that perhaps her ideas were not thought out clearly enough before they were spoken. I was not, to be truthful, looking to expend a lot of energy on deciphering contradictions. I had read too much about the importance of consistency and reproducibility in research that I didn't want to touch these sorts of contradictions, as if they might somehow 'invalidate' my study. I asked myself what I could possibly do with 'data' that appeared to contradict itself. In fact, even when I decided to pay closer attention to her voiced conflicts and tensions with regard to knowing science, I did so from a perspective of a researcher interested in the concept of participant reflexivity. In other words, I wanted to use her words as fodder for the reflexive process - asking Evelyn to reflect on her words and what it meant to her that these were some of her perceptions on the knowing of science. I think I was expecting her to be hit by the apparent contradictions and then attempt to clarify herself. In fact, we were both hit by the apparent contradictions but, in speaking about them further, we realized that they were real. Evelyn really felt the tension of being pulled ***between*** *the poles of pressure to know* ***and*** *support for not knowing. We even realized that the pull of each pole often originated with the same groups of people. Parents often place great expectations on their child's teacher to know science and to be able to impart this knowledge to their child, yet at the same time they often understand from their own experiences (and even share with their children) just how 'difficult' science is.*

Apple (1995) refers to certain ideological contradictions that dominate our society as "structurally generated contradictions" (p. 107); for example, in our society we are urged to work hard, yet at the same time we are given messages that real satisfaction in life is not in work but in leisure. These conflicts and tensions are experienced and reproduced at various levels in schools as well. Apple (1995) states that "… in any real situation there will be elements of resistance, of struggle and contradiction, all of which will act against the abstract determination of the real life experiences of human actors" (p. 85). While there is increasing emphasis on the school as an agent in the reproduction of dominant ideologies, Apple (1995) warns against believing fully in deterministic models of reproduction. He states that "social and cultural life is simply too complex to be caught by totally deterministic models" (p. 85) and that "we must contest a particular assumption—that of passivity" (p. 87). Hegemonic ideologies— with their commonsense practices and universal truths— are still subject to internal inconsistencies. Perhaps it is one of these inconsistencies that should be the foothold for resistance. So, while accepting that the contradictions are real and present, this did not help me to understand them— not until I began to read more about resistance.

A few pages ago in this chapter, I gave examples of what I have termed 'self-preservation' resistance, in which my participants alluded to keeping science at a distance because they were not very good at it or did not have the capability to do it. Their resistance to science seemed to rest mostly on issues of confidence, feelings of comfort, and perceptions of ability. In many respects then, a resistance about 'self-preservation' is really a resistance embedded in notions of identity and agency—finding pathways which are perceived to be open for personal resistance and (re)action.

> Agency is never freedom from discursive constitution of self but the capacity to recognize that constitution and to resist, subvert, and change the discourses themselves through which one is being constituted. It is the freedom to recognize multiple readings such that no discursive practice, or positioning within it by powerful others, can capture and control one's identity. (Davies, 2000, p. 67)

I believe, however, that my conversations with Evelyn point to a different form of resistance (other than self-preservation). The statements about there being a lot of *pressure* to know science and, at the same time, a lot of *support* for *not* knowing science, within and among other preservice teachers, is what I will call ***'cohesive resistance'***. This form of resistance appears to almost approve of a stance to not take on the role of learning and teaching mathematics and science (but particularly science). Its formation may be 'justified' or explained through a reflection on personal role models or an expression of how science is often portrayed, but its *cohesiveness* seems to create a cultural form of resistance among elementary preservice teachers.

While it is not a conscious decision to buy into such a hegemonic ideology, it

> **Did you know that...?**
>
> Giroux (1981) warns us not to speak of a depoliticized notion of culture, divorced from notions of class, power, and ideology, because then culture becomes an empty social science category. He uses culture to suggest "not the existence of one 'whole social process' happily produced by all members of society, but different layers of meanings and practices mediated by the inequitable distribution of wealth and power" (p.148). He states that a view of multiplicity of cultures is not accurate because the situation is more like the existence of "a dominant culture (with its own contradictions, of course) and of the existence of minority cultures, all mediated by considerations of power and control" (p. 148).

does lead to a reproduction of the 'norms' of the dominant culture. The very acceptance of this form of resistance is a form of ideological reproduction.

> [T]he very contradictions that students live out in their day to day lives may end up supporting the institutions and ideologies that they seem to oppose yet offer a terrain for action at the same time. (Apple, 1995, p. 84)

There is a paradox here; one that I am having difficulty expressing. Rejecting the worlds/cultures of school mathematics and science by reason that they do not relate to our lives (an action that may be viewed as serving the interests of the dominant culture in deciding what knowledge is of most worth), could work to further reify the division between school mathematics and science and our lives. By refusing to 'take it on' and relinquishing to its 'difficult nature', we are further emphasizing the division between 'those who can' and 'those who can't', meaning that the same small subset of people will buy into it (whether useful to them or not) and the larger subset will be deemed incapable. This is an admirable form of resistance (to just not take it on) but it is misdirected and disorganized. Instead of focussing on the root cause for the cynicism and distrust, the resistance is aimed at alleviating feelings of discomfort by not buying into its high status and reproduction. Would it not be more productive to be critical of the ideology we are being asked to reproduce, and not just *resist* reproduction?

Could it be that my participants' life experiences directly contradict the daily experiences of school science and mathematics? Does it not make sense then that clear and obvious contradictions between school mathematics/science and their lives might lead to rejection? If learners (preservice teachers, etc.) are told (over and over) how important mathematics and science are to their everyday lives and to their futures, yet they consistently see contradictions in this—that, in fact, school mathematics and science appear to have no direct effect or bearing on their everyday lives at all—then is it not a reasonable outcome for these learners to become distrusting and cynical of these messages, and thereby reject and resist the importance of mathematics and science? And in doing so, will they also look for others to support this resistance— to condone its rejection?

"The seeds of reproduction lie in this very rejection."
(Apple, 1995, p. 90)

the existence
of resistance
is key

If it exists,
then it can be (re)formed to work toward emancipatory ends.

If it is not even present,
if one simply accepts all that comes their way as natural and necessary,
then there is no place to begin.

I believe that 'cohesive resistance' contains many elements for potential action and agency. The strength and/or insight to resist in the first place are necessary first steps.

Giroux (1981) views ideology as a form of social reconstruction rather than a set of illusions or lies. He defines ideology as "a set of beliefs, values, and social practices that contain oppositional assumptions about varying elements of social reality, i.e., society, economics, authority, human nature, politics, etc." (p. 148). "Hegemony is an ideology that defines the limits of discourse in a society by positing specific ideas and social relationships as natural, permanent, rational, and universal" (p. 148). He further states that "ideologies become hegemonic when they are institutionalized by the dominant society" because it is at this point that "they are stripped of their oppositional power and serve to legitimize existing institutional arrangements and social practices" (p. 148).

"Through its structured silence about what ought to be, hegemonic ideology elevates 'common sense' to a universal truth." (Giroux, 1981, p. 149)

An ideology itself— in the form of beliefs, values, and practices— is not hegemonic, but it is the unquestioning acceptance of the ideology of the dominant culture as natural and common sense that constitutes hegemony. The desire is not to strip preservice teachers of oppositional power, but to refocus and redirect it.

In his discussion on how Enlightenment goals went astray, Bernstein (1976) states that

> [S]cience and technology have become powerful and invidious tools for manipulation, repression, and domination, rather than the way to enlightenment and freedom. The common reaction to this increasingly ominous state of affairs is either to fluctuate between forms of pessimistic despair that claim we are caught in a process which has its own logic, over which we can exert no control, or to attempt, through some form of romantic protest, to reject or escape from our scientific and technological civilization. It is a measure of our growing sense of impotence that even to label this situation a crisis has degenerated into a cliché. (p. 176)

The first form of "pessimistic despair" described in the above quote seems to be a form of false consciousness, while the second form has been illustrated in this text through examples of Mike Michael's (1996) work.

As the seminar member said to me after my presentation (see page 156), "people find interesting ways to resist". I was not initially receptive to these ways in my conversations. I favoured blaming false consciousness, thinking that a research goal of 'true' consciousness would be all at once enlightening and empowering for my participants.

The glaring issues of power, and what I perceived as its resulting oppression, had blinded me to consider that students do not passively accept all that comes their way. I wanted to believe in 'power as sovereignty':

> Power as sovereignty often creates a dichotomous world in which there are the oppressor and oppressed, thus producing a dualism whose effect is to define particular social groups as unified entities. The dualism of oppressor/oppressed loses sight of the productive qualities of power, that is, how systems of knowledge generate principles through which action and participation are constructed. (Popkewitz, 1999, p. 6)

But, power is not as simple as *one has* and *one does not have*. Foucault (1990) writes: "There is not, on one side, a discourse of power, and opposite it, another discourse that runs counter to it. Discourses are tactical elements or blocks operating in the field of force relations" (pp. 101-102). According to Foucault, the tactics form comprehensive systems, composed of multiple relations all operating in a sphere which is both constituted and constituting, changing moment by moment, and not traceable to any one institution or structure. "Power is everywhere; not because it embraces everything, but because it comes from everywhere." (p. 93)

FOCUS on...

Michel Foucault (1926-1984). "The provocative and influential French thinker [who] challenged the fundamental Western liberal assumptions, originating in the Enlightenment, that history is the objective story of society in progress [...] Foucault's work strikes hard at science [saying that] all scientific investigation and social discussion (discourses) about human nature, cultural activity, and social history, emerged from social agents engaged in processes of social control, that is, of power." (Horus, 1998)

Very concerned about the social relations of power and the power of social relations, it was often easier to decipher from his work, what he thought power *is not*. One such description is that "power is not what is now fashionably called 'cultural hegemony'." (Horus, 2000)

> For Foucault, power is not a monolithic force that appears in the same guises throughout all times and places. Instead, power has a genealogical history and is to be understood differently depending on place, location, and theoretical perspective... For Foucault, power is capillary, flowing throughout the social body and not simply emanating from on high. (Dimitriadis & Kamberelis, 2006, pp. 114-115)

> Every social practice, including that of being schooled, calls forth the body. And every body is burdened by history's way of seeing it. … the bodies of the students became sites of struggle in which dominant symbols and metaphors designed to regulate the body were resisted much the same way a transplanted organ by a foreign donor is often rejected by a body's tissues. Resistance also became a way of gaining power, celebrating pleasure, and fighting oppression in the lived historicity of the moment and the concrete materiality of the classroom. (Giroux, 1989, pp. 194-195)

It seems to me, however, that one form of power speaks louder than the others—almost as if one is '**P**ower' and the other forms are '**p**ower'. Perhaps I am confusing *strength* of P/power with *organization* of P/power. If the tactics of exercising power through the various forms of resistance are focussed (subconsciously) in conflicting directions and operate on a level that is not yet fully realized or acknowledged, then perhaps their overall effect in the sphere of force relations is somehow lessened. In any case, I will return to issues of **p/P**ower and its implications for this study in the next chapter, when I ponder the images of mathematics and science knowing formed through, and by the power of, epistemological lenses. In this section, I am concerned with exploring how power might be present in the forms of resistance offered by my participant preservice teachers. As Foucault (1990) reminds me: "Where there is power, there is resistance, and yet, or rather consequently, this resistance is never in a position of exteriority in relation to power" (p. 95).

Inside (my) Research

Reflection and absorption working together to perpetuate the dominant culture...

Notions of colour theory:

"What is the colour of the sky in your world?" I have always liked this (sarcastic and rhetorical) question because it gives a clear message that what a person is saying or how she/he is acting seems so out of the ordinary that she/he must be from a different world. But what is the ordinary anyway? Too often the dominant culture is the reference point for ordinary, and all else is assessed relative to this 'norm'. All situations are **coloured** by the cultures that are brought to the situations (Ted Aoki, July 10, 2000, University of Lethbridge Summer Institute). We are often unaware of this unconscious dimension, but we *are* **colouring** with the culture that we bring. It is important to become aware that this is true for *all* situations. Unfortunately, absorption represents a **colouring** of/by the dominant culture, so that the **colour** of the sky in *your* world is based on the **colour** of the sky in the world of the dominant culture. We absorb and reflect exactly what we are told.

I see a red shirt because the pigments in the shirt have absorbed all colours of the visible spectrum *except* red, which is reflected to my eyes. The properties of the material determine which wavelengths of light will be absorbed and which will be reflected. Absorption and reflection have nothing to do with the light, unless of course the light is somehow filtered before hitting the object/material/shirt.

What does it mean then to say that we colour each situation with the culture that we bring to the situation? It is necessary to talk about both absorption and reflection in order to understand colour theory.

Absorption... knowledge that we take in; our minds are capable of soaking it up like a sponge but, when squeezed, we will give back exactly what was taken in, unchanged by any property of the sponge. Absorbing the colours of the dominant culture means eventually giving back the knowledge(s) of the dominant culture, unchanged or unmodified by our own culture. In fact, we don't feel that any interaction of our own culture with the knowledge/situation is even acceptable or legitimate.

Reflection... knowledge that we refuse to accept; we want no part of it and we will resist it by reflecting it back to the sender. Reflecting the colour(s) of the dominant culture means that we refuse to partake in the knowledge of the dominant culture.

TRANSMISSION AND REFRACTION

"…schools do not transmit culture, instead they play a fundamental role in reproducing the dominant culture." (Giroux, 1981, p.149)

Did you know that…?

Transparent materials do more than just transmit light. They also **bend** or refract, its path. Refraction, or the bending of light, occurs because light travels at different speeds in different materials. Not only does the amount of bending depend on the types of materials through which light passes, but it is also a function of the **angle** at which light hits the boundary surface.

> To see that all knowledge is a construction and that truth is a matter of the context in which it is embedded is to greatly expand the possibilities of how to think about anything, even those things we consider to be the most elementary and obvious. Theories become not truth but models for approximating experience… (Belenky et al., 1986, p. 138)

A constructed way of knowing involves a transformation of how one thinks about and embodies knowing, using the self as an instrument to understand, question, and transform knowledge and perceptions of knowing. It is, of course, perceptions of knowing that I have been predominantly interested in during the course of this research.

"Constructivists seek to stretch the outer boundaries of their consciousness… by imagining themselves inside the new poem or person or idea that they want to come to know and understand." (Belenky et al., 1986, p. 141)

One of the reasons I use the word perceptions when speaking/writing about preservice teachers' epistemological views is to highlight the tacit dimension of knowing. "Perception… appears as the most impoverished form of tacit knowing." (Polanyi, 1966, p. 7) One of the problems with the perpetuation of objective idea(l)s of knowledge is that perception is considered irrelevant and external to knowing. In objective views, there is a schism between knower and known which leaves no room for the personal participation and perception of the knower. But Polanyi (1966) states that "into every act of knowing there enters a passionate contribution of the person knowing what is being known" (p. xiv). He proposes that knowing is a type of tacit integration of clues into meaning and that all knowledge is an act of tacit knowing. Gelwick (1977) describes the fundamental structure of all knowing, according to Polanyi's thought, as involving three centers of knowing, or what he refers to as a "triad" (p. 63). This triad consists of a problem (or focal target), the clues (or the pieces of information of which we have subsidiary awareness), and the person. Polanyi's structure for knowing clearly emphasizes the active and necessary involvement of a person who is capable of creating meaning by integrating subsidiary clues into the focal target. Moving from the clues to meaning is an achievement of the person, done within the body, making knowing a form of "indwelling".

"Because our body is involved in the perception of objects, it participates thereby in our knowing of all other things outside." (Polanyi, 1966, p. 29)

The process by which a person makes meaning through the integration of subsidiary clues to the focal target is frequently inarticulable and is a unique achievement of the person. In the perpetuated idea(l)s of objective scientific knowledge, however, there is usually no room made for this person.

> The objective ideal of knowledge is propagated and renewed through the philosophies of knowledge that continue to foster schisms between mind and body, between reason and experience, between facts and values, or, in short, between the knower and the known. This epistemological dualism is the citadel of the objective ideal of knowledge. (Gelwick, 1977, p. 141)

Now, I turn to the metaphorical relationship. Refraction is a phenomenon that occurs when light is transmitted through a material, such that the properties of the material determine how the light from an object will be changed or bent as it passes through the material and how, ultimately, images will be formed. In a metaphorical sense, I believe the notion of refraction can be compared with Polanyi's move from clues to meaning that must be done by the person in all acts of knowing. The refraction of light is like indwelling, where the meaning-making process occurs within the body. Just as there is an important phenomenon occurring in the spaces between object and image formation, there is an important phenomenon occurring between the particulars and a meaningful whole in coming to know. This process can be explained through the theory of tacit knowing— one that accounts for the personal participation of the knower in all acts of knowing.

Perhaps Polanyi's most fundamental description of the tacit dimension of human knowledge is summarized in saying, "*we can know more than we can tell*" (Polanyi, 1966, p. 4). As educators, caught in cycles of measuring and assessing, we often emphasize that we know *only* what we can tell, ignoring the tacit dimensions. An illustration of how personal participation, or the third element of the "triad", is neglected in the objective idea(l) of knowledge is in the portrayal of knowledge being "out there", waiting to be discovered. Preservice teachers have generally not considered the perspective that that which is apparently "out there" is still a social construction. Curriculum learning outcomes/objectives play a very deceptive role in portraying knowledge 'construction' more as knowledge 'acquisition' or 'commodification'. There is a perpetuated idea(l) of producing identical knowers. In other words, notions of constructivism in school mathematics and science teaching and learning are only superficially engaged— they are used in an attempt to reach the point where every learner is allowed to bend and makes sense of their own knowledge, but *only to the extent that it ends up looking exactly like everyone else's.*

The metaphor of refraction helps illustrate the absurdity of such an overall learning goal—that of producing identical knowers. Light interacts differently with different materials and, as a result, the images formed are also very different. Now imagine the expectation that identical images should be formed regardless of the material through which light passes. It is as absurd an expectation as requiring that all learners— regardless of their individual and collective 'refractive' possibilities— reach a final endpoint in which all knowledge images (including what is important to know and why) are identical.

"We are in a world of knowledge without genuine knowers, a situation of absurdity." (Gelwick, 1977, p. 95)

I propose that 'refraction', or a transformation of views of knowing, is only possible through a critical reflection on the way knowledge is portrayed as unproblematic and as simply a matter of direct transmission and acquisition. To view knowledge as problematic "posits a tentative view of knowledge as socially constructed, subject to political, economic, social, and cultural forces, and contingent upon communities of discourse, relations of power, and social change" (Britzman, 1991, p. 48).

"... the art of knowing is seen to involve an intentional change of being." (Polanyi, 1966, p. 64)

It is with/in this topic of image formation that I continue the metaphor of exploring what happens when light hits different surfaces. As light 'passes through the material' (indwelling), it is transformed (or bent) according to both the properties of the material and the positioning of the object. The next chapter will explore this transformation through the metaphorical phenomenon of refraction, by exploring how images are (*or could be*) formed through lenses.

What happens when light hits different surfaces? REVIEW

Unscramble each of the clue words.

Copy the letters in the numbered cells to other cells with the same number.

BRTISPOANO					1						
CETRELONFI			3								
TIRNOECARF	13		12			6		17			
LTHIG					7						
GEACONNIR									5		
ESNIELC						15	2				
TESIRESNAC	10										
HEEGONMY						8	19				
RULCOO			4	18							
LUERCTU				16			11				
NEANEMSILGS			14	9							

1	2	3	4	5	6	7		8	9		10	11	12	13	14	15	16	17	18	19

How are images formed in lenses?

We desire what we've lost
as well as what we've never known.
We look and look with our bodies,
dissolve into places and loves, into earth and light.
We look into present moments
to enter what has passed there before us,
to take apart a moment like a landscape...
(Michaels, 1995, p. 183)

CHAPTER FOUR

HOW ARE IMAGES FORMED IN LENSES?

preVIEW...

In the first three chapters, I explored common experiences of elementary preservice teachers with respect to their perceptions of what it means to know (in) mathematics and science. In this chapter, I begin by presenting a poetic re-view of participant voices in order to set the context for a discussion on how school mathematics and science might be re-imag(in)ed. For this discussion, I draw on the light metaphor of how images are formed in lenses. In exploring the possibilities for understanding different ways of knowing (in) mathematics and science, traditional views of objects, images, and lenses are disrupted and deconstructed. The chapter draws on the analogies of object positioning, lens properties, and image characteristics in order to highlight tensions, or aporias, en route to imag(in)ing other possibilities for knowing (in) mathematics and science.

[see **Voice Re-view** on next two pages]

One aspect of my research conversations that has not really been discussed yet is a perspective on how my participants view themselves if they *do* like mathematics and/or science and if they feel they *are* 'good at' these subjects. I touched on this briefly, when speaking about the satisfaction of finding answers and 'understanding' bonding. For the most part, however, my participants' images of themselves as mathematics and science knowers did not generally indicate a level of comfort and satisfaction.

In keeping with an image of mathematics and science as being removed from their lives and capabilities, my participants occasionally made comments about not fitting in (with the norm) if one *is* 'good at' mathematics and science. To me, such comments appeared as subtle indications that their images of mathematics and science knowing were wrapped up in pre-determined ideas of who is supposed to "do" them. This highlights another contradictory and exclusionary aspect of mathematics and science images. A double-edged sword is apparent: mathematics and science are only for the very few who are "smart" and "capable" because the subjects are difficult, but then if one *is* "successful" at mathematics and science, she/he automatically falls into this "smart", not-so-normal category. It seems as if people want to explain

> We didn't do anything in grade eight.
> Before that I loved math, I loved it.
> I don't know...
> it's kind of geeky, eh? (laugh)
> (Helen, November 1999)

away or make excuses for why they are successful at (or like) mathematics and/or science, sending the message that they are embarrassed to be different.

Voice Re-View

Why would someone say they weren't good at science? Probably because they didn't get somebody's right answer.

It's written down somewhere, someone's documented the answers, someone has access to the information. You just have to know how to get it.

I used to work from the answer all the time. Like, get the answer from the back of the book then... once you have that answer, you almost have the confidence to do the question.

I think if you start going into middle years and high school, and you came up with the wrong answer, they'd be like 'yeah, you don't know anything.'

I accept it as a fact actually
[that with science] it is more likely I'll be wrong.

If you're not given room to doubt, then how are you going to question?

Don't question, just do it.

Nobody understood that I had to know why or else I couldn't do it.

I don't know if other people just don't *think* about why you're supposed to be doing this or if they have an *understanding* that I don't have.

Like I said, if I didn't have to take science, I wouldn't.

I always felt like she was teaching to the male students in the class.

After a while, you don't even raise your hand anymore 'cause you know you're not going to get picked.

Our chemistry teacher told us that the only real thing that the girls could get from chemistry would be the chemical equation for diamond.

I accept it as a fact actually
[that with science] it is more likely I'll be wrong.

They gave you a handout, you memorize it quick the night before and you go spit that out and forget it the next day. I'm very good at [definitions]. If I had to have a science test, I'd want it to be that... 'cause I can rattle it off for you.

[Science is] not something that you really understand off a page. At least I can't.

It was all memorization for me. Now that I'm realizing it doesn't have to be that way, I really am kind of resentful.

Like I said, if I didn't have to take science, I wouldn't.

Science was always 'fill in the blank space here', right or wrong.

Maybe we put too much emphasis on the *word* and the things that make it difficult. Like science words are *scary*.

The word science gave me a knot in my stomach. I have this fear of the idea of not understanding.

I accept it as a fact actually
[that with science] it is more likely I'll be wrong.

So, some of these kids will be coming up with things, and showing me, and I'll have no idea how it works. That scares me... 'cause I won't be able to understand what they're talking about.

It's like this little sub-culture and I'm on the outside... at the fence; you know, [I] can see it but I don't know how to get to it.

We're taught to use our experience as knowledge and yet it's disqualified if it doesn't fit in. Why?

That can't be science... yeah, it makes sense, I get that... it can't be *real* science.

Like I said, if I didn't have to take science, I wouldn't.

I accept it as a fact actually
[that with science] it is more likely I'll be wrong.
Like I said, if I didn't have to take science, I wouldn't.

If you *were* good at math and you're a girl,
or if you wanted to do math and you're a girl,
you were kind of a freak...
'girls shouldn't be doing math, it's too hard'
or I don't know why...
but the guys *had* to do math... they had to do all the work...
even if they didn't get good marks— which a lot of them didn't—
they still had to be *in* the math.
But the girls,
'well, if you want to just take general math, that's ok'
I never did, but a lot of the girls would go to the counsellor
and say
'I'm having trouble with math'
and they'd say
'why don't you consider general math?'
But [with] guys, it would be like
'try harder'

(Helen, April 2000)

Aside from the ideas expressed in our conversation about the images of mathematics and science knowing, I also asked my participants a slightly different question: I asked them to describe the characteristics of science. Shortly after asking the question, however, I realized that I was probably going to be misinterpreted. After all, the question **'what is science?'** is commonly encountered in elementary and middle school classes. Often, the teacher/curriculum is not overly concerned with what the students' perceptions *really are* about science; the question is usually rhetorical, asked as a lead-in to the first chapter of the text, where the answer to this rather philosophical question is laid out in a very unambiguous manner. For example, this is the exchange between Evelyn and I...

<u>Kathy's Question:</u>
If you were to describe to someone what science is, your image of it, what characteristics would you give it?

<u>Evelyn's Answer:</u>
I think to me, science is *why* things work, *how* they work. It's a lot about interconnectedness, how they fit together. Sort of like you have this big picture where you see how this relates to this—but, at the same time, also, what *is* this? How does it break down into *its* components?

Kathy's Grade 9 Science Notes

Without wanting to disregard Evelyn's response, I wanted to further clarify what I meant specifically by my question. Since this part of our conversation occurred after already speaking with her for more than two hours about her experiences in, and perceptions of, **science**, I somehow felt that her 'answer' above represented more of an agreed-upon description of science instead of one that reflects her own personal experience of science. I thought that perhaps the best way to clarify my intended question would be to change the wording a bit and ask her to describe the 'person science' as she 'knows' or experiences it— as if she was describing it to someone else who has never 'met' science.

Portions of her response have been included already in this text (in chapter two) because it helped illustrate what I called the *particle theory of knowledge*. Her chosen metaphor of the rebus is quite powerful and portrays, I think, the contradictions that come to the surface when trying to answer: 'what is science?' During that same conversation about the rebus, Evelyn pondered the 'person science', and stated:

In one way, to me science is like...
it's like when somebody says to you

'oh, there's this person you *have* to meet, they're great,
you're going to love them!'

And you meet them, and you think

'well, actually, I don't really like them.'

I know I'm really *supposed* to enjoy science;
I'm supposed to find it interesting and be passionate about it.
But I think, 'well, *ok*... that was interesting.'

I think you get sort of stuck in an idea of what science *is.*
And for me, it's something that's difficult for me so I don't really move that far past that. I mean, it's great if I understand something but...

I think it's a disadvantage for me not to *be* excited about it.
If I'm excited about something, I will learn it and find out about it.
But, really, I don't think I've ever been that excited about science.

Science for me is a lot about being unsure.

(Evelyn, December 1999)

In another conversation with Ursula, I asked her the same question. She responded:

I've never met a math or science yet that I like.
(Ursula, January, 2000)

Both Sylvia and Harriet reflected on and articulated their images of science in the following conversation piece:

December 1999

H: It's evil... no, it's not evil (laugh). It's confusing.

S: It's very abstract for me. It's something out there and I'm not sure really how to get to it, but I know it's out there. When I think of science and math, it's just always out there... I *know* it's there but I don't know really how to get there, what would be the easiest way to get there to *be* this mathematician or scientist that I always see these pictures of and I have these pictures in the back of my brain. And still—even though I'm learning how to be a teacher of science—I still see bald men, with grey hair, standing up in lab coats and thick glasses.

K: Someone you can't identify with?

S: Noooo... and that will never go out of my head... even though I'm learning that's not really what science is, that's still there. And somehow I'm still striving to get to that goal even though I know I'm not supposed to anymore, like I don't have to. But yet, I don't know, I feel I won't be a complete teacher of science until I'm somewhere close to there.

K: And what does the process along the way represent for you?

S: Oh, it's like a winding path, with forked roads and you're not really sure where to go. Like, it's not clear and it's not nice.

H: I agree with [Sylvia]. For me—math and science—I don't know, makes me shake my head and... Like, I think I'm a smart enough individual. Maybe that's why it frustrates me because other things I do, school-wise, I'm confident I can do, gung ho. But with math and science, when you mention it, I get sweaty palms, and it's this big issue.

In articulating her images of **science**, Sylvia draws upon metaphorical language that clearly speaks of her feelings of exclusion and confusion. She will never *be* a bald man, wearing thick glasses, in a lab coat— so does this mean that she feels like an imposter? A phony? A winding path with forked roads, not being sure which way to go, illustrates an ever-present confusion for her as she attempts her journey toward being a science teacher.

Later in that same conversation depicted above, Sylvia and Harriet reflected on their experiences with experiential learning in science in the teacher education program courses. The following comments are helpful in understanding how the image of 'science as a puzzle' continues to make them feel uncomfortable with their knowing in science:

Harriet: It's a subject where you can have fun and not be learning a damn thing.

Sylvia: The thing we're learning is so much experiential learning, child-centered, dah, dah, dah, dah, dah... doing all these experiments, which is great, but still somehow from behind the scenes we have to build up these concepts which is more of a direct way of teaching. And I'm fine with the experiential: 'let's mess around with this'. That's good for me, but getting behind the scenes, to *teach* them the concepts, like why does this happen. There's always the concepts there... you *can't* go through the whole year just doing experiential learning in science—as much as I would *love* to do that—there's still underlying concepts that you have to learn, that are there and you just have to learn them; things you just have to know and to get to that point where I'm comfortable teaching that—almost like a lab coat: 'this is how you do this' thing—that's scary for me... just getting behind the scenes—'cause I don't even know really what those concepts are and I'm not really sure how they relate and how am I suppose to teach that, if that makes sense.

Harriet: It's like I've got all these little jigsaw puzzle pieces that I know should fit—I've been *trying* for *so* long to make them mesh but I can't and that's the frustrating part. After the four months [of science education classes], we're supposed to come out and go 'yeah, yeah, yeah' and we, for the most part, are like 'yeah, but...'

I find it particularly interesting how Sylvia has come to love "messing around" with science, but clearly sees it as being a very different part of science teaching and learning from the underlying concepts that "you just have to learn." I think the tension voiced here is a critical aspect of the lived experiences of these preservice teachers. Both Sylvia and Harriet (and other participants) clearly distinguish between two versions of science—the playful and the serious. The playful aspect is like doing a puzzle just for the fun of it, without having to think about the underlying meaning of the activity (or even whether there *is* one); the activity is fun and creates lots of enthusiasm, but it is not necessarily associated with learning anything. But eventually, one has to get to the *serious* science. Herein lie the definitions, terminology, and "underlying concepts" that are most often associated with the *learning* of science (probably because they are most often associated with the **testing** of the learning of science). It's as if each puzzle piece has an important meaning embedded in it and a particular connection to be made but, to these preservice teachers, it no longer seems like a fun activity to attempt to figure out just what those meanings and connections are. The tension rests in "meshing" these two versions of science; in other words, to unearth and cultivate the deeply hidden context and connections between the *playful* and the *serious*. Playful and serious are seen as binary opposites and the gap between the two seems ever expansive to Sylvia and Harriet. Because of this, they feel inept at the serious science, as if they have no conceptual understanding of science whatsoever. Is it possible to explore these gaps—to create a to-and-fro movement between/within these binary opposites so that it is not **either** playful **or** serious, but **both** playful **and** serious?

For the most part, my participants' references to their experiences in mathematics and science blended together in our conversations; both subjects often represented confusion and exclusion for them. Evelyn was one person who had quite different experiences with, and attitudes toward, mathematics and science. She described the 'person' math quite differently:

> I have success with math.
> When I uncover a piece in math, I say
> 'Oh, that makes sense; that fits in with everything else I know.'
>
> (Evelyn, December 1999)

One of the critical goals of my research has been to try to understand how and why women's experiences of knowing (in) mathematics and science have been explored predominantly by focusing on the women themselves as knowers (Belenky et al., 1986). Initiatives abound that attempt to make women 'fit' mathematics/science, perhaps because it is easier to focus on changing women than on changing mathematics/science. There are a few important pedagogical initiatives that attempt to make science 'fit' women (see, for example, Davis & Steiger, 1993; Rosser, 1997; Roychoudhury et al., 1995). My criticism of several of these approaches, however, is that I believe they still focus on the problem of women not fitting in. The goal is often to take the same school science curriculum as always (along with the standard answer-oriented assessment procedures) and dress it up a bit so that women will find it slightly more appealing to learn. Such initiatives often risk essentializing women in that they propose things like co-operative groups and other 'female-friendly' environments. Women are still viewed through a fixed lens of what it means to know (in) mathematics and science. I believe that putting feminist theory into feminist practice requires a different tactic—one that asks different questions differently; that is, by changing the focus of the questioning.

As I write in this metaphorical language of images, lenses and focusing, I am reminded of some very memorable high school optics exercises…

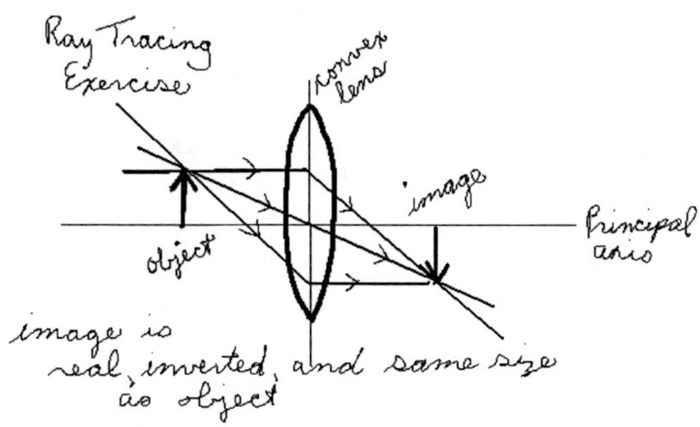

Similarly (or metaphorically), I think knowers often set themselves up relative to a lens, obtaining an image of themselves as knowers. This image of themselves as knowers has particular characteristics (smaller or larger than themselves, real or virtual, upright or inverted, and positioned relative to the lens). Working with the definition of an image as "an apparent object" is very interesting, since it helps place perspective on the entire notion of images.

Did you know that..?

Each part of a visible **OBJECT** emits light (either its own or reflected) in all directions. Diverging rays of light entering the eye causes the eye to "see" a point of light at the location where the diverging rays of light would meet if projected backwards. If light rays diverge directly from the points comprising an object, the eye will reconstruct the points and see the object directly. An object, then, can be defined as a visible substance from which light diverges and passes directly into the eye.

If the light from an object is reflected, refracted, or in some way deviated, it will diverge from points at different locations of the corresponding points comprising the object. The eye then sees the object at a different location in space from that actually occupied by the object. Such an apparent object is called an **IMAGE**.

(Bruenn, 1968, p. 231)

im·age (ĭm′ĭj) *noun*

1. A reproduction of the form of a person or an object, especially a sculptured likeness.
2. ***Physics.* An optically formed duplicate, counterpart, or other representative reproduction of an object, especially an optical reproduction of an object formed by a lens or mirror**.
3. One that closely or exactly resembles another; a double: *He is the image of his uncle.*
4. **a.** The opinion or concept of something that is held by the public. **b.** The character projected to the public, as by a person or an institution, especially as interpreted by the mass media.
5. A personification of something specified: *That child is the image of good health.*
6. **A mental picture of something not real or present**.

(AHDEL, 1992)

The image of oneself in a lens is not fixed; it depends on the characteristics of the lens (concave or convex, thin or thick, single or combined, etc.), and also the characteristics of the 'self' relative to the lens (including positioning and attitude).

Quote of the day...

We never did *real things*.
We'd read about things
but it wouldn't *connect* to anything
I knew about.
(Evelyn, October 1999)

In this chapter, I am devoting much time to the exploration of optics terms because I believe the optical metaphor is helpful in illustrating how images of knowing (in) mathematics and science are **both** problematic **and**, at the same time, open to many new possibilities. (Not to mention the fact that these *playful* connections of optics terms to everyday language, like image, power, attitude, etc. were never really made in the *serious* learning of high school physics!) At this point, I will explore the metaphorical problem(atic)s associated with image formation. I will do this by identifying what I am calling **panoptical aporias**.

> **pan·op·tic** (pàn-òp¹tîk) also **pan·op·ti·cal** (-tî-kel) *adjective*
> Including everything visible in one view.
> [From Greek *panoptos*, seen of all : *pan-*, pan- + *optos*, visible.]
> (AHDEL, 1992)

> **aporia:** [*fr. Gk., difficulty, perplexity*]
> a problem or difficulty arising from an awareness of opposing or incompatible views on the same theoretic matter;
> *esp*: one giving rise to philosophically systematic doubt.
> (Webster's, 1976)

Skovsmose (2005) adds to this definition by identifying an aporia as "a 'puzzle', 'question for discussion' or 'state of perplexity'. An aporia refers to a difficulty, which remains open" (p. 120). I refer to these metaphorical problematics as aporias because I believe that spaces of possibility can be envisioned by pulling the problem, or state of perplexity, wide open— by disrupting these ruptures.

The first aporia I will discuss is that of *complicity*. The lens through which my participants' images are shaped and positioned is the lens of mathematics and science knowing AS THEY HAVE DESCRIBED IT (in all its (not) glory!). The voices of my participants throughout this text, and especially in this chapter, illustrate some of the ways in which my participants view mathematics and science knowing— a winding path, bald men in lab coats, puzzles, an uninteresting person. If this is the lens through which preservice teachers attempt to locate an image of themselves as mathematics and science knowers then it is no wonder that they slip into the binary oppositions of right/wrong, **good at / not good at**, etc. The lens of mathematics and science knowing (as they have described it) is restrictive, exclusionary, difficult, disconnected, and completely impersonal. And, most detrimental of all, the situa-

tion is portrayed as if this lens of knowing is unchangeable, untouchable and detached from the knowers. Are preservice teachers, and learners in general, complicit in accepting this image formation process?

The second aporia, to be found in the context of locating one's image as a knower in/of mathematics and science, is that of *objectification*. An image is only an apparent object; we must not confuse it with the object itself. In setting up this metaphor, however, it appears that the knower is being objectified. Even without this metaphor, however, the knower is frequently viewed as the object— the object of research, the object of initiatives to increase participation in mathematics/science, the object of why schooling fails. By not questioning one's personal participation (or lack thereof) in the acts of knowing, knowers *are* being objectified (and objectifying themselves). To draw out the parallels between knowers as objects and the objects of optical image formation, imagine positioning and describing oneself as an object with characteristics of…

> size, type (real or virtual), position (distance from lens), attitude (upright or inverted)

relative to this (perceived) fixed lens, with properties…

> concave or convex, thin or thick, single or combined,

where the lens becomes 'fixed' according to one's experiences of/in knowing (influenced or 'fixed' by marks, teachers, and other experiences). In such a scenario, where the knowing object and the lens are viewed as 'fixed', all one can do is hope for the best! What one obtains is an image of oneself as a mathematics and science knower.

Reflexions...

In her discussion on "five aporias toward a curriculum in distress" (p. 4), Patti Lather (1999) states:

> *Between the no longer and the not yet, the task is to produce different knowledge and produce knowledge differently, layering complexity, foregrounding problems, thinking outside easy intelligibility and transparent understanding. The goal is to move educational research in many different directions in the hope that more interesting and useful ways of knowing will emerge. (pp. 7-8)*

Lather is speaking of educational research paradigms. When I read this quote, I hear the same words in my mind but with 'educational research' replaced by 'mathematics knowing' or 'science knowing'. It's all the same, isn't it? Is this what my research is about? Daily, I confront aporias of legitimation, interpretation, difference, objectivity, and complicity (and others such as representa-

tion and voice). In my mind, these aporias are the gaps or crevices in/through which there exists a hope of grabbing hold and tearing wide open, as one way to move with/in and beyond ideological boundaries. Maybe the 'no longer' in mathematics/science education signifies the search for one Truth, for value-free and completely objective knowledge. Then again, maybe I am referring to the 'still is'. As long as the space between 'no longer' and 'not yet' is ambiguous, messy and slippery, will mathematics/science knowers let go of the 'no longer but still is'?

Now that two (of the several) possible aporias have been introduced, what can be done with them? As mentioned, I think that possibilities exist for these aporias to be deconstructed and disrupted. The question to ask is: What can be done in the metaphorical physical space between object and image, if the lens occupies a fixed position, and the image has distinct characteristics relative to this fixed lens? Nothing can be done unless we look at the experimental optics set-up differently— that is, disrupting the aporias by re/imag(in)ing the object and the lens. I have two suggestions:

change object...

To address the aporia of objectification, I suggest a radical departure from, and a complete re-positioning of, the optics arrangement. Instead of placing women (or knowers, in general) as the objects— who are then viewed through the lens to locate an image— place mathematics and science (knowing) as the objects; mathematics and science are the objects we are coming to know. How can they be re/imag(in)ed and re/positioned?

change lens...

Once the objects have been changed to mathematics and science (knowing), imagine viewing these new objects through different lenses—lenses representing different ways of knowing and different ways of being a knower. It appears that the notion of having different lenses available has never been an option for my participants. Perhaps the idea of changing the lens through which they could view these objects (as opposed to subjects! ☺) would be similar to mathematics and science (knowing) "coming to them" in different ways.

Kathy in conversation with Helen *(December 1999)*

Kathy: People are often saying to me, 'what do you think this research is going to do? I mean, science is science, right? It's there, whether it's men or women who are doing it, whether people think they know it or they don't know it... IT'S SCIENCE.' But you could say the same about anything if you choose to make it very dogmatic, or rule-based. If I said to you that you can't write a story in a Language Arts class because you can't tell me everything about all of the parts of speech, you'd think I was ridiculous, right? But the image of science isn't the same.

Helen: It's very interesting. My mind's going like a hundred miles an hour.

Kathy: What are you thinking?

Helen: I'm just... I'm making all sorts of connections really, really fast with everything that I've ever known in my whole life. We talked in our [women in education] class about how science is always portrayed as the be-all-and-end-all of knowledge.

Like, if you know about science, well you know *stuff*.

We talked about other ways of knowing... like how you know someone's mad just by the way they look... that's knowledge. I think it's because of the way science is portrayed. I don't know by who, or when that started...it must have been a long time ago. I think it's a Western thing.

Inside (my) Research *Two of my colleagues and I recently submitted a paper to a particular journal on women and 'other' minorities, relating to science and engineering fields. I was actually fairly conventional in how I wrote my part of the paper—I left out my poems, my text word art, and most of my other performative features that I enjoy play/writing with. I included a few minor (non-APA) inter/disruptions because I thought a journal devoted to writing/reading women's voices might be supportive of difference. The reviewer's comments, however, did not really surprise me.*

A research participant quote, written with intentional <Enter>s for poetic effect, rather than one long extended line of words with lifeless voice...
received the written comment: "write like all other APA formatted quotes".

A sentence using parentheses to denote either and/or both as with "knowing (in) science"...
was circled as if to suggest that I must have forgotten to carry out the proper spelling and grammar check on this page.

A participant quote presented inside a lightly shaded text box, to set it apart from all those other theoretically-validating quotes from people out of context that I don't even know...
attracted the reviewer's attention as s/he wrote (but perhaps did not think) outside the box: "remove shading".

So
there I was
writing inside the box of shade
thinking outside the box of their shadows
... and do you want to guess whose shadows prevailed?

(Nolan, 2005, p. 135)

Reflexions...

I suppose that I have become saturated with the expression 'ways of knowing'. It seems to be lacking something. My perception of this expression is that the focus of research still seems to be on figuring out different ways to know the same thing, or different ways to get to that same answer. It seems almost patronizing at times. Too much research starts from the position(s) of women, which immediately has the effect of sending signals that women and their epistemological positions need careful understanding, and even fixing.

*Instead of using the expression 'ways of knowing' for my discussion that follows, I'd like to coin a new phrase: something which takes emphasis **off** 'fixing' the knower and **on** rethinking the object(s) to be known. What about 'ways of sciencing' or 'science viewing' or 'science imag(in)ing'... yeah, that could work— looking at different ways of creating/locating images of science means exploring how we position science relative to the lens of knowing so that science can imag(in)e itself differently! This idea reiterates my philosophy of the need to ask different questions and ask questions differently. Starting with the voices of my participants and their experiences, I would like to try to (re)present some of the possibilities for a reconstruction of imag(in)ing science and imag(in)ing mathematics.*

*Hmmmm. Speaking of lenses and images, I seem to be converging to the language of science and diverging from the language of mathematics. I certainly do not mean to imply that spaces of re-imag(in)ing are limited to science knowing. While some imag(in)ings may focus more on one subject area more than the other, the **mathematics and science imag(in)ings** that follow have implicit connections for, and reflections on, both subjects (objects?) of study.*

MATHEMATICS AND SCIENCE IMAG(IN)ING NARRATIVES

In Physics... the only thing I remember from that class was that Marie Curie died of cancer from radiation and her daughter died of cancer from radiation... like why do I know *that* but I couldn't tell you the first thing about radiation? Or that Einstein was a rebel and almost got kicked out...? (Ursula, October 1999)

"The very notion of telling a story presupposes at least an audience and at best an active community of meaning-makers."
(Povey & Burton, 2004, p. 46)

Burton (1996) writes about narratives in mathematics in particular, but she also delineates some very important general characteristics of the nature of narratives in learning. A few of these characteristics are (p. 30):

> Narratives are personal, told by and often about the experience of the narrator vested with the meaning that the narrator chooses to garner…
>
> Narratives are used to pose, explore, and respond to *What if?* type questions.
>
> Narratives may be communicated in many different forms, in speech, in writing and/or pictorially.
>
> Narrative is participatory, involving a community in telling and responding to a story.
>
> Narrative is *a*, possibly *the*, way to explore the meaning of experience.

Burton (1996) proposes "that a narrative approach to mathematics and its pedagogy is consistent with a view of mathematics as being socially derived and with the understanding of mathematics as being socially negotiable" (p. 32). She emphasizes the importance of telling the socio-cultural story of/in mathematics so that it may be possible to imagine moving away from a presentation of mathematics 'content' as fixed facts and outcomes. In their discussion of author/ity as a way of knowing, Povey & Burton (2004) state that:

> The notion of mathematics as narrative helps us to 'see' the authors of mathematics within a given community. This human meaning-making has been expunged from the accounts of mathematics that appear in standard textx; the contents are then portrayed in classrooms as authorless, as independent of time and place and as that which learners can only come to know by reference to external authority. (p. 47)

> Narrative... presupposes different perspectives, different strategies, different conclusions, allowing knowledge conflicts to emerge naturally and themselves to become the focus of reflection in the classroom.
> (Burton, 1996, p. 33)

> In our science [education] class, we had the resource 'Keepers of the Land' introduced to us... [it's] an aboriginal story about a woman who was angry about how the people were treating her and [so she] became part of the land. I don't remember exactly what the story was about but... after, there was a science activity based on westernized views of science that linked the two... and I thought that was neat... the story could make the link. (Evelyn, February 2000)

In relation to the subject of science, Kuhn (1996) writes about paradigm shifts within textbooks of 'normal science', presented as a rewriting that leaves no trace of previous theories and renders invisible the revolutions that called for the 'rewriting' in the first place. Threadgold (1997) referred to this as "the erasure of the intertextual histories of scientific texts" (p. 18). I have two major criticisms of the (scientific) approach to the re-writing and re-presentation of science. Both of these criticisms are embedded within a critical statement made by Threadgold (1997): "the performance or making of science... is a profoundly embodied and disciplined process, subject to all kinds of policy, institutional, private and power relationships, and it is always narrative, dialogic and rhetorical" (pp. 21-22). My point here, too, is that science **is** subject to power relationships and science **is** narrative. And yet, it is seldom portrayed as having any characteristics of narrative...

narrative: story tale account description plot storyline sequence of events

Perhaps it is believed that if the storied nature of science is emphasized then there is a risk of science appearing less rigorous, maybe even subjective or fictional. I wonder, would that be a *negative* effect of science imag(in)ing narratives? Again, we are presented with a binary opposition in need of deconstruction: theory/narrative. Theory is privileged over narrative, as if theory were present, just waiting to be discovered, whereas narrative is a re-presented 'tale' from someone's perspective. Theory is presented as disembodied; narrative as embodied. Of course, in presenting science as disembodied, the nature of the performing bodies can be denied. Threadgold (1997) points out how the "embodied masculine subject of science" (p. 21) disappears from the making (or *poiesis*) of scientific theory in the acts of transforming science from performance narrative to written reports of the performance. Not only is the binary opposition of theory/narrative imposed here, but, in doing so, all traces of the performers vanish. We are left only with the finished written piece, disembodied and lifeless.

MATHEMATICS AND SCIENCE IMAG(IN)ING THE EVERYDAY

Helen illustrates how she is beginning to place science, not herself, as the object in her realization of how physics could be imag(in)ed through the everyday world:

> For me, with physics...
> well, you have to take the math and the science and put them together,
> that was way too much for me.
> But... because I thought that you'd have to know
> a *whole bunch of stuff* about science
> and learn *enormous amounts*
> about a whole bunch of different little things,
> I thought I'd never be able to do it.
> I just said, 'no, that's it, I can't do it'.
> But now... [I'm realizing that]
> physics is an all-encompassing thing,
> it's everything,
> I didn't *know* that.
> I think if I would have known that physics is ... pretty much the whole world,
> I might have been more comfortable going in.
>
> I didn't know that physics could explain the 'WHYs' of everything.
> (Helen, December 1999)

> **Quote of the day...**
>
> we don't see it as science
> we see it as everyday life...
> it's too easy, so it must be wrong
> (Harriet, December 1999)

Roth & Barton (2004) describe a classical approach to science and scientific literacy as one where "science is a pure subject, often taught in special physically separated rooms, unsullied by common sense, aesthetics, economics, politics, or other characteristics of everyday life" (p. 74). In direct opposition to this image, they advocate for "conceptualizing scientific literacy as a feature of collective praxis" (p. 74); that is, science and scientific literacy understood as a community practice and not "pre-packaged in books or delivered to students away from the lived-in world" (p. 3).

MATHEMATICS AND SCIENCE IMAG(IN)ING METONYMIC SPACES OF DOUBLING

I have already written considerably on the topic of metonymic spaces of doubling in the first two chapters of this book. It has been a challenge in my research (for both me and my participants) to imag(in)e learning mathematics and science without focusing on answers. Ursula asked me, "... but how can you teach math *without* worrying about answers?" And, in one of our discussion groups, Elsie emphasized the importance of answers when she said that, for her, not having an answer for someone would be "identifying a weakness" in her. As long as school mathematics and science curricula focus on 'learning outcomes', and encourage objective assessment of these outcomes, then it will be difficult to imag(in)e ambiguous and messy spaces in the teaching and learning of mathematics and science. Mathematicians and scientists live ambiguous and messy spaces daily, so why is this not reflected more in school curriculum?

Goos, Galbraith, and Renshaw (2004) speak to mathematics teachers (and students) in reminding them that the job of a teacher is not to remove the frustration, ambiguity and messiness of learning:

> ... perhaps the most difficult task confronting the teacher is to learn what *not* to do, that is, to resist the urge to do the mathematics for the students, and to let them grapple directly with ideas in what might appear to be a messy and inefficient fashion. (p. 113)

In addition, writing on the lack of messiness present in certain, pre-determined outcomes, Gordon (2006) states:

> ... the quest for certainty in education has contributed to the narrowing of the curriculum, the proliferation of instructional practices aimed at teaching to the test, and the marginalization of important dimensions of our beings like developing the imagination and a sense of mystery. (p. 17)

Davis, Sumara, and Kieren (1996) are critical of "the convictions that learning outcomes should be pre-specified and that, once such objectives are selected, learning sequences can be devised to ensure that they are effectively achieved" (p. 152). In writing about an embodied curriculum that focuses on possibilities, Davis et al. (1996) propose "a 'good enough' theory of curriculum, where the goal is to uncover not what is ideal, but what is possible" (p. 166). These authors explain that if students and teachers embrace the possibilities of a 'good enough' theory of curriculum then "we must be prepared to face a much more ambiguous set of curricular-relations which would be uncomfortable in a world that expects clear limits and boundaries" (p. 166).

MATHEMATICS AND SCIENCE IMAG(IN)ING NEW BOUNDARIES

In proposing a reconstruction of science education, Clay (1996) requires us "to question the value of a school curriculum that is based on the assumption that the ways of knowing can be compartmentalized into five, seven or ten subjects and within that framework, that some subjects are intrinsically 'superior' to others" (p. 192). Burton's (1996) response to the complexities involved in (mathematical) literacy for the twenty-first century is: "Far from wanting to encourage young people to keep discipline areas discrete, I would prefer a curriculum which grew out of a problem-perspective using all appropriate knowledge bases" (p. 38).

teaching out of the text,
not making it real,
and not crossing boundaries between subjects—
to me, that's against my religion.
(Roberta, December 1999)

> The existence of subject hierarchies, and furthermore, the content within those subjects that constitutes 'valid knowledge', is itself circumscribed by patriarchy and its dominant values. (Clay, 1996, p. 192)

Earlier in this book, I wrote about my desires for mathematics and science integration, but the language and implications of 'integration' just didn't satisfy me. Then I wrote about interdisciplinary and problem-based learning, but that still seemed somehow lacking in substance. There are many initiatives introducing problem (or project)-based approaches to teaching and learning mathematics (Morgan, Watson & Tikly, 2004; NCTM, 2003; Stevens, 2000) and science (Goodnough & Cashion, 2003; Grow & Plucker, 2003; Keys & Bryan, 2001), but few of these challenge the traditional 'detached' and apolitical views that are perpetuated in the teaching and learning of school mathematics and science. In order for mathematics and science to imag(in)e and embrace new boundaries, it is necessary to make explicit and critical links between mathematics and science and the social, political, ethical, and economic contexts that are part of the learners' worlds (within local communities and in society at large). Research in mathematics and science education that embraces such critical contexts and perspectives is beginning to have a significant impact on the *theories* of teaching and learning, but much remains to be done at the levels of classroom practice. (See, for example, Barton, 2003; de Freitas & Nolan, *in press*; Gutstein, 2006; Roth & Désautels, 2002; Skovsmose, 2005; Valero & Zevenbergen, 2004). With such extended and reconfigured boundaries, questioning the rhetoric and consequence of actions (and inactions) form a new optic for what we should know and do in school mathematics and science.

MATHEMATICS AND SCIENCE IMAG(IN)ING ENACTIVISM

In writing about the enactment of engaged pedagogy in a science classroom, Meyer (1998) states that one pedagogical direction she has taken is to view "science as participation with natural phenomena by enacting possibilities, interruptions, and interpretations, playing with variable relations and using our bodies, voices, memories and imaginations…[such that] knowing and doing are implicated in each other and the split between body and mind is blurred" (p. 466).

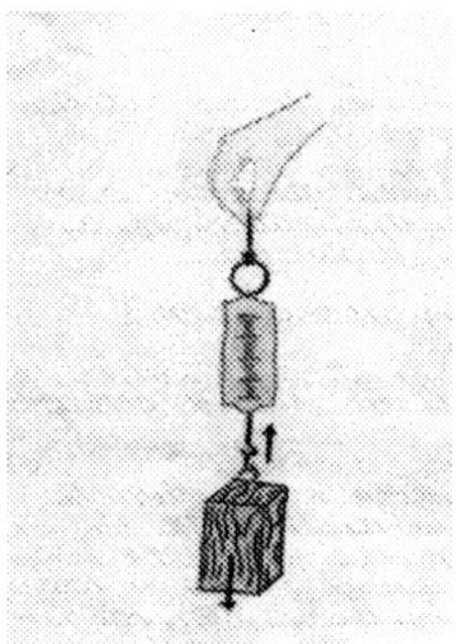

In my first reading of enactivism in the science classroom (Fels & Meyer, 1997; Meyer, 1998), I understood it to be just another way of advocating "hands-on" science. Now, I understand it to be more about "bodies-on" science; in other words, using enactivism as a way to involve our own bodies in conSCIENCEness. My science experiences taught me that science is a mental endeavour (with the mind) and occasionally a manual endeavour (with the hands), but I certainly never experienced it as a corporeal endeavour (with the body).

Several of my participants also called for such an embodied conSCIENCEness, after having experienced a very limited view of science…

history is repeating itself
there's too much memorization…
especially science

science should be
experiencing
living
breathing
doing

Not just memorizing
and photocopying

The natural blending of drama and storytelling in the science classroom, in order to explore alternative ways of knowing, doing, and being with science (conSCIENCEness) is what Fels & Meyer (1997) call performative inquiry.

> In performative inquiry, we realize our journey through performance and then map our pathways through discussion, reflection, remembering, writing, re-

imagining. There are no final answers, no final destination. There is only process, journeying, exploring, possibility. (Fels & Meyer, 1997, p. 77)

"... our body is the ultimate instrument for all our external knowledge."
(Polanyi, 1966, p. 15)

MATHEMATICS AND SCIENCE IMAG(IN)ING OPEN-MEMBERSHIP

Evelyn: What I'm thinking is that I probably haven't tried very hard to understand science either and it's probably like a rebus: if you work at it, you'll understand it and ... maybe you'll look at it and think 'oh, yeah'.

It's just another way of thinking... like if someone gives you a riddle...
If you're not used to them, you don't understand but, once you get used to them ...
'what has four eyes and never sees?' You think: 'oh, it's Mississippi'

So maybe that's the way it is with science too...
it's just a different way of thinking that you have to train your mind to get into doing and once it's there, it's fine.
It's just wanting to be there or getting there... the process and the desire.

Kathy: Have you given any more thought to the idea of turning that around and thinking that maybe it's *not* always your job to try to work hard to figure out what science is doing but that maybe there should be work on the other end to let you *see* it in a different way?

Evelyn: Yeah. I think... I used conspiracy theory (laugh). I think there hasn't been a lot of effort made by scientists to help people or students learn about science in other ways than, you know, simply formulas and facts. There hasn't been a lot of encouragement for people to *go* beyond basic thought. There's an idea of science as:

"we're a closed group, a private club, maximum membership 50 people;
sorry, you're just not going to be able to apply *this* year"

and I think there hasn't been very much support in alternative views of science.

I don't think there's a lot of challenging on the ideas that 'only *this* is science'. In social studies curriculum, [there are] always special interest groups, saying 'what about us?' But with science there don't seem to be many other voices, and the other voices are often portrayed as... sort of crackpots even.

I think science has been really limited by *its* idea of what science is and who can *do* it.

IMAGinE
the
IMAGE
through the lenses of mathematics and science imag(in)ing
other ways of knowing

I think I'm not a strong science person but I don't think it's my fault necessarily. I think that... I *know* that I had bad experiences, but I don't think my experiences were all that different than the experiences of people who *did* do well. So then I'm confused. I know for sure that mine weren't good experiences, but how were they any different than the experiences of the girl who sat beside me and did just fine? They were the same experiences... maybe we were just different learners. I don't know. (Ursula, October 1999)

Maybe, in fact, we *are* all different learners and this is why we need to learn in different ways. Another quote by Ursula earlier in this chapter suggested that she remembered the stories about Marie Curie and Albert Einstein, but she 'learned' little else in her physics class. Does this mean we should question her cognitive capacities for physics? Of course, this is often exactly what does happen. If Ursula is placed as an objectified knower, giving her the fixed lens through which to view her image of a science knower, then the result is likely that she sees a smaller, inverted, virtual image that makes her feel very much out of place in learning physics. Is it possible to allow her to choose her own lens through which to view physics— through which she will obtain the largest, most real, and best-positioned image of knowing and learning?

To speak of a large, real, and best-positioned image is to once again speak about "engaging the Will" (Noddings & Shore, 1984). One *will* want to learn something that is real and interesting in one's life—Evelyn said this, Helen said this. Interest often comes with having choices—Ursula said this, Evelyn said this. Answers to questions of what counts as knowledge and what knowledge is of most worth in mathematics and science have been 'fixed' and unchallenged for a very long time now and appear closed to re-imag(in)ing. 'Learning' therefore usually means eventually getting to the point where one 'knows' (or can, at least, repeat) this decided upon set of knowledge(s), after having traveled the single, pre-determined path for reaching that point. Davis et al. (1996) prefer to portray learning as a much more complex, and much less linear, phenomenon:

> Learning should not be understood in terms of a sequence of actions, but in terms of an ongoing *structural dance*—a complex choreography—of events which, even in retrospect, cannot be fully disentangled and understood, let alone reproduced. (p. 153)

Returning to the quote by Ursula (on the previous page), it is worth emphasizing that no one can, of course, speak for the girl sitting next to Ursula. Ursula's hypothesis about how people can have the same experiences yet 'learn' differently is crucial, I think, to understanding how learning depends on the lens through which one views what it means to know something. Most times in school, we have no choice of lens because the final image (of knowing) is made identical for everyone.

Perhaps the girl sitting next to Ursula was me—I did not *choose* a lens of science knowing through which I saw only definitions, formulas, and carefully coloured notebooks, but this was my experience. It just so happens that I was the type of person to 'master' or 'learn' whatever was in my field of view of required knowing, simply because I was told to and I was capable of it. Perhaps if I had not been capable of 'learning' this way then I would have been more critical of the lens at the time. I doubt it, however. Like several of my participants, I probably would have interpreted a failure to 'learn' this way as an inherent weakness of my own. Perhaps I would have claimed, "I am just not very good at it".

That conversation with Ursula continued by reflecting on why *her* science experiences resulted in 'failure' while the girl sitting next to her became a science 'success'.

Kathy: I don't know actually. I think it's the same reason why people look at the *numbers*… trying to increase the numbers of women going into science because then that will prove maybe that science is ok the way it is.

Ursula: That's only part of it though. You can have all the women you want doing [science] but if *nothing changes* you're still going to lose how many of us in the dust because we *can't* get it? It's just a symbolic number of people. I'm not saying that nobody should do it, but if it's just the token people who are going to do it because someone gave it to them as a challenge… it's not saying *science* is any different. It's not giving you any different way to *look* at science. It's just saying 'well look, so-and-so did it, it must be ok; you should try it'. It's not going to make you any less frustrated when you're the person who's confused.

A Moment in HERstory...

Ursula Franklin (Sept. 16, 1921 -). Born in Munich, Germany, Ursula acquired her Ph.D. in experimental physics from Technical University in Berlin. Presently, she is Professor Emeritus of Metallurgy and Materials Science at the University of Toronto. In May, 1995, Dr. Franklin gave an address entitled *Looking Forward, Looking Back* at a conference at the University of New Brunswick. In this address, she looked back on her own experiences of being the sole woman in an engineering class (where her professors asked her on the first day if she was in the wrong classroom) and looked forward to the situation for women in engineering. Ursula identifies that one typical approach to 'the number problem of women engineers' in this patriarchal field and society is that of "weight-lifting for girls".

> Though I understand full well the need to encourage young women to enter engineering and to support them personally in every possible way, acculturating women into engineering and hardening them against the chilly climate may change the problematic aspects of the culture of engineering less than one might think. (Franklin, 1995)

A second approach, which goes well beyond the number focus, is a systemic one. Ursula describes this as being more of a question of structural, institutional and cultural changes—directing attention at the patriarchal structures of hierarchy and power—with a focus on making "engineering fit for women, rather than women fit for engineering" (Franklin, 1995).

The word 'imagine', embodying the word 'image', makes me imagine what our images of mathematics and science knowing would look like through so many different possibilities for types of lenses. And what might happen if we combine lenses in this reimag(in)ing? There may be some reluctance, in reimag(in)ing curriculum, to encourage *too* many lenses for fear of some sort of…

Did you know...?

"...the failure of a lens to give a perfect image are known as **aberrations**. They have nothing to do with the composition of a lens or the smoothness of its surfaces, but arise naturally from the geometry of image formation. They may be reduced to an acceptable level, but, in general, the reduction of one type of aberration tends to increase another" (Jones & Childers, 1993, p. 625).

ab·er·ra·tion (àb´e-râ′shen) *noun*

1. A deviation from the proper or expected course.
2. A departure from the normal or typical: *events that were aberrations from the norm.*
3. *Psychology.* A disorder or abnormal alteration in one's mental state.
4. **a.** A defect of focus, such as blurring in an image. **b.** An imperfect image caused by a physical defect in an optical element, as in a lens.

chromatic aberration *noun*

Color distortion in an image produced by a lens, caused by the inability of the lens to bring the various colors of light to focus at a single point.
(AHDEL, 1992)

In locating an image through a lens (or combination of lenses), the first thing scientists want to do is to reduce or eliminate the aberration, in order to make the image one unified entity. I'm using the metaphor of lenses and aberrations to explore how pre-determined learning outcomes attempt to ensure that all images of knowing look the same and that there is no "deviation from the proper or expected course." I believe curriculum pre-determined learning outcomes try to eliminate all blurred or ambiguous images of mathematics and science knowing. In some sense, this appears analogous to trying to eliminate difference. By presenting the knowing of mathematics and science as clear and unambiguous, the concept of difference is (hegemonically) eliminated. The philosophy at work, it seems, is that if very clear and distinct images of knowing are presented and 'normailized' then this would exclude different knowers and knowing differently.

I think, however, that we should be careful of equating difference with an aberration, and thus work to eliminate it, as western science often does. Instead, we should focus on and embrace difference. Difference colours the situations in which we actively participate. Why not embrace chromatic aberration?

> The survival of productive difference requires that we put all claims for intellectual hegemony in their proper place—that we understand that such claims

> are, by their very nature, political rather than scientific. (Keller, 1995, p. 179)

Joan Scott (1997) proposes that a critical feminist position on difference involves two simultaneous actions: "The first is the systematic criticism of the operations of categorical difference, the exposure of the kinds of exclusions and inclusions—the hierarchies—it constructs, and a refusal of their ultimate 'truth'" (p. 768). The second action addresses the refusal of the equality/difference dualism. She suggests that it is not in women's best interest for feminists to seek equality that rests on sameness or identity but "an equality that rests on differences—differences that confound, disrupt, and render ambiguous the meaning of any fixed binary opposition" (p. 768).

> *Sameness and difference collide over and over again as impulses toward multiplicity and national good continually vie for prominence... (LeCourt, 2004, p. 24)*

Inside (my) Research

I want to reflect for a moment on Derrida's concept of différance and the significance of it to my research. Différance calls to mind, simultaneously, the spatial quality of differing and the temporal quality of deferring in meaning. It enters into my research in a critique of the standard Western metaphysical binary oppositions, where presence is privileged over absence, light over dark, etc. I have been attempting to disrupt or deconstruct these, and many other binary oppositions, 'present' in Western thought and 'absent' with/in standard critiques of knowledge in science. "Différance is actively disruptive. Language, thought and meaning aren't to be allowed the comfort of their daily routines" (Collins & Mayblin, 1996, p. 77). Setting "in play a movement of undecidability" (p. 77), différance gives me the encouragement I need to imagine a deconstructive reading of the binary oppositions.

> In its elaboration of a critique of the metaphysical forces that structure and smother différance in every text, a deconstructive reading thus assumes... [t]hat an inquiry that attempts to study an object by means of that very object is open to certain analyzable aberrations (this pertains to virtually all important investigations: the self analyzing itself, man studying man, thought thinking about thought, language speaking about language, etc.). (Johnson, translator's introduction to Derrida, 1972, p. xvi).

The process of deconstruction, then, does not 'produce' another meaning which is to replace (or directly overturn) the traditional binary opposition of Western metaphysics. Instead, meaning is deferred, and undecidability plays in the spaces between either/or. This is also where Derrida's concept of différance comes into 'play' in my research. Play in these spaces is what I have been referring to as **both/and** metonymical spaces of double.

To disrupt and render ambiguous the fixed meanings of binary oppositions has been a key interest of mine in this research. The overall detriment of dichotomous thinking has been illustrated with preservice teachers' perceptions of themselves as knowers. There appears to be the perception that in order to obtain the 'right and true' image of oneself as a knowing object in the lens of mathematics and science knowing, it is important to ignore difference, to reduce aberrations. In doing so, however, agency is removed from learners in the discourse(s) of science. "The interests which a discourse serves may be very far from those which it appears, at first sight, to represent" (Weedon, 1987, p. 122-123). Discursive relations of science are always in fluctuation and meaning is (should be?) deferred. Hekman (1990) attempts to bring some clarification to Derrida's use and meaning of the term différance when she suggests "defin[ing] it as that which always escapes, is deferred in the attempt to define absolute knowledge as presence" (p. 25). Of course, to fix meaning is a logocentric view and is very much the western view of science, making it appear as if discourse is not based in context and historical perspective. By removing ambiguity and multiple interpretations of meaning, the lens through which we view mathematics and science knowing appears fixed and unchangeable. These are the hidden effects of power.

Power: in optics, a measure of the amount by which a lens will deviate light rays. The unit of power is the **dioptre**, which is equal to the reciprocal of the focal length in metres. By convention, the power of a converging (convex) lens is positive and that of a diverging (or concave) lens negative.

$$P = 1 / f$$

As can be seen from the mathematical relationship above, the power of a lens is inversely proportional to the focal length of that lens (i.e., the shorter the focal length, the weaker the power).

(Hutchinson, 1995, pp. 168, 326)

"Discursive practices are embedded in material power relations which also require transformation for change to be realized" (Weedon, 1987, p. 106).

In continuing the lens metaphor… to examine the deviation of light rays is to study the property of refraction of light. In the previous chapter (on what happens when light hits different surfaces), one of the possible paths for light to follow when incident on a surface is transmission *through* the surface. When this occurs, light's path is almost always bent, or refracted. In a metaphorical manner, I believe this property of the refraction of light parallels notions of changing, or transforming, images of knowing in mathematics and science. By this I mean that it is important to encourage open discussions on a transformation in thinking about, and critically questioning, the discourse(s) of mathematics and science—questions of what counts as knowledge, what knowledge is of most worth, and who decides. When light travels from an object through a lens, it is the phenomenon of refraction that enables one to know how and where the light rays will converge to form an image. The size, location, and attitude of the image will depend on the properties of the object, as well as the properties of the lens (one of which, by the way, is its power). If we can imag(in)e replacing knowers as objects with mathematics and science as the objects, and encouraging multiple lenses, then I believe there are possibilities for a re-organization and deconstruction of power relations. Weedon (1987) does not seem overly optimistic about such deconstructive possibilities with/in the discourses of science, stating:

> The institutional investment in science which confirms the status quo is massive. Feminists do not have the resources or the institutional positions and backing to make much impact on the discursive hierarchy of existing writing and research, backed, as it is, by capitalist and patriarchal interests. (p. 132)

Personally, I am slightly more optimistic, especially after reading Lewis's (1993) words on discourse, transformation, and power:

> … discourse is an action taken upon the world. These actions may be transformative or they may be tenaciously preservative of the status quo; whatever the case, discourse is socially negotiated through power. (Lewis, 1993, p. 114)

I also take heart in the words of my participants.

Evelyn asks about conspiracy…

> I was wondering if—
> and maybe it doesn't exist so much in high school, but in university—
> there's some sort of conspiracy to keep people *out* of math and science.
> I know [in first year] math,
> lots of people fail it and the profs don't change the curriculum.
> They keep it the same.
> It's an expected thing, that people will fail math.
> (Evelyn, October 1999)

Evelyn also comments: "science has been really limited by *its* idea of what science is and who can *do* it". Ursula thinks that the push to have more women *do* science is really just about having "a symbolic number" and that "if it's just a token [number of] people who are going to do it because someone gave it to them as a challenge… it's not saying *science* is any different… it's not giving you any different way to *look* at science." And then there is Helen's surprise at realizing, after all these years of choosing *not* to study physics, that it could have helped explained some of the WHYs of the world for her; she expresses her annoyance at having missed out on that. All of these voiced recognitions and frustrations give me hope, in this research, that my participants are at

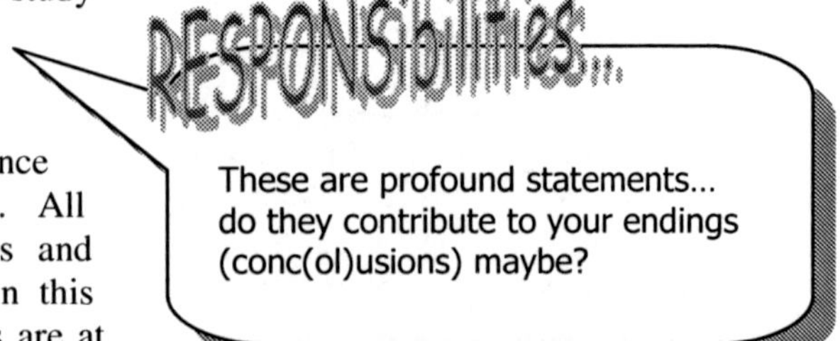

least pondering the idea that being "tenaciously preservative of the status quo" can be challenged because "discourse is an action taken upon the world" (Lewis, 1993, p. 114). What about Harriet's "sweaty palms" at the mention of science and Evelyn's realization that she is at a disadvantage to not be excited about science —can these preservice teachers continue to live with/in these tensions and contradictions in their daily teaching lives, without acting on them?

Do the voiced reflections of elementary preservice teachers in this study suggest that a critical questioning and re-examination of lived experiences is underway? In metaphorical terms, are my research participants beginning a journey of "seeing" things differently? In many ways, their wor(l)ds are optimistic ones because they represent, to me, a questioning of things outside of themselves. Through these reflections on their lived experiences, there is a promising glimpse of a recognition of their agency to act. A re-imag(in)ing of the metaphorical lens arrangement—changing object and changing lens—is a necessary first step toward a range of possible transformations in images of knowing and learning (in) mathematics and science. Questioning who has knowledge, what counts as knowledge and why, and what knowledge is of most worth challenge hegemonic discourses by recognizing that they are the political, historical, and contextual questions which give meaning and shape to power-knowledge relations.

> "Meaning is always political. It is located in the social networks of power/knowledge relations which give society its current form. Not all areas of discourse are equally significant in the hierarchy of power/knowledge relations but no discursive practice is outside them." (Weedon, 1987, p. 138)

How are images formed? REVIEW

Fallen Phrase Puzzle

A phrase has been created and the letters of the phrase have "fallen" below the grid. Each letter shown appears in the same column, but just below where it should be. The solver must put the letters back in the grid and rebuild the phrase.

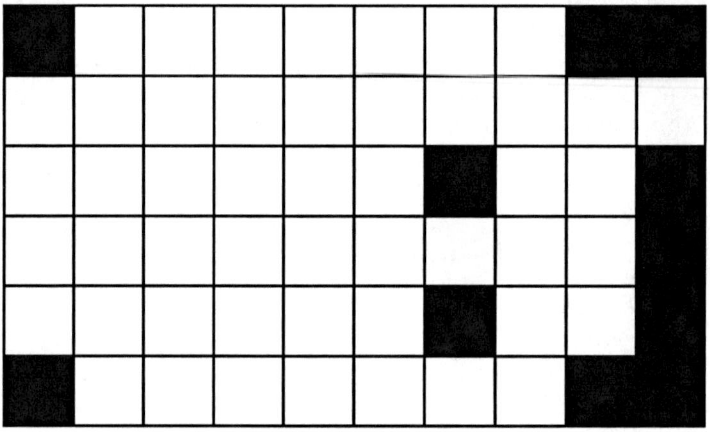

	K	F	M	L	R		N		
	I	S	I	W	G		I		
I	E	N	O	A	N	I	N	F	
D	E	C	E	E	I	E	G	T	
L	T	I	S	E	F	C	E	N	
R	S	N	F	E	S	N	O	E	S

CONC(OL)LUSIONS

Part I: A Kaleidoscopic (Re)View

So if I explore a certain experience of incompleteness, of deep-down craving,

oh, I can tell you exactly what I *don't* want
but to tell you what I *want*...

I'm not suggesting it's the only matrix for living your life. It's simply the one I know. (Lee, 1995, p. 38)

We would mostly just sit...
and not draw on the desks

It was something that you study and you try to get the right *answers*

I'd better just acknowledge the question,

It always *feels* like a race to an *answer*

and let it breathe. (Lee, 1995, p. 44)

If you get *lost in the middle*...
it *is* a problem

How would you *measure* it?

They're *scared*
I'm *terrified*

Insist on the importance of faithful, yet critical and necessarily partial, stories of women's experiences and material circumstances for producing critical understandings of how patriarchal

The only real thing that the girls in the class could get
from chemistry would be the chemical equation for diamond

Teachers would say

oppression is legitimated by hegemonic epistemic values. (Code, 1995, p. 181)

if the *girls* didn't *want* to do it,
they could *just sit* there
You know,
after a while,
you *don't* even raise your hand *anymore*

Science is not something that
you can really *understand*
off a page

And *underneath* it all would be a *rebus*

If you're not given any *room to doubt*
Then how are you going to *question* it?

My poor secret shadow
how long have you walked

obscured in the dark of my shade?
You were born right behind me

Once you have the *answer*,
you almost have the *confidence* to do the question

like a dog on a chain
watching all the mistakes
I have made.

"Dog on a Chain"
(Fearing, 2000, track 6)

So you don't have to *understand* anything to do that.
You just do the action. That's not hard for me.
I can fill the *same thing* in over and over again...

I think it's because we did it *over and over again*,
and we just kept doing it.

After you finish the first two and a half years of your [university] program
then you can stop and think about the '*whys*'
but before then you just *have to get the facts*.

I think science has been *really limited* by *its* idea
of what science is
and *who can do it*.

Resonance is a form of clarity.
(Zwicky, 1995, p. 74)

I don't think it's so much what I actually *know* about it...
but I *don't want* to say that I [know it] because I *don't want* to have to *do* it.

For this passion is the doorway for imagination;

It wasn't real *learning*.

here is the possibility of looking at things

It was all *memorization*.

as if they could be otherwise.

I really *hope* I can do things *differently*.

(Greene, 1995, p. 16)

CONC(OL)LUSIONS

Part II: See(k)ing Reasons

Shadowed by light,
Knowing by heart

When we imagine a reason to change
In our hearts we know
All is not dark in shadow

The presentation of this research text as a physical science textbook, divided into chapters on light, has presented a kaleidoscope of counternarratives.

> Such counternarratives are… quintessentially "little stories"—the little stories of those individuals and groups whose knowledges and histories have been marginalized, excluded, subjugated or forgotten in the telling of official narratives. (Peters & Lankshear, 1996, p. 2)

These (counter)narratives (which I prefer to call "vital" stories, rather than "little" stories) are tied together by the metaphorical thread of the properties of light, but also held apart by the tensions and contradictions with/in this critical epistemological exploration.

Allow me to return, for a moment, to the constellation of inquiry that was presented in the Introduction to this book. I cannot resist— after having written this kaleidoscopic text of counternarratives— asking the question: "Have we really seen the beautiful things that the sky holds?" (McLaren, 1994, p. 265)

One of the questions presented in that constellation was to ask how one can construct counternarratives to the 'problem of women in science'. The wording of 'problem of women in science' has been used to reflect the everyday hegemony of focussing on women as a problem, as something that needs fixing. It has been my experience that there are several official narratives associated with the problem of women in science, with most of them focussing on the need to address imbalance and inequity by enticing more women to study science. The implication in such an official narrative is that if the gender numbers somehow balance, then the 'problem' will be alleviated. No matter how often I have (re)presented the stories and voices of my participants with respect to their experiences of gendered stereotypes and different curricular expectations, the official narratives still linger in thoughts and emerge in conversations as confirmation that hegemonic spaces thrive. If, for example, I am giving a presentation on this research, there will always be—without exception— someone in the audience who will ask me if it is not in fact the case that the numbers of women in science are actually much higher these days. I do not think we have really seen the beautiful things that the sky holds.

The many vital stories of preservice teachers' voices, challenging official everyday narratives with/in the language of light and knowing, may help us imagine a critical journey. Through these vital stories, I hope that journeys have been started; but they are in fact personal journeys of change. As Griffin (1995, p. 6) writes:

> The change proceeds in so many directions
> I cannot be its narrator.
> I am instead immersed in
> the atmosphere of this motion myself,
> my own thoughts changing even as I write.

I write in the language of the properties of light, but the process of actually 'seeing' is a personal journey. The vital stories are intended to present thought-provoking challenges to the official stories, hopefully opening eyes and awakening hearts.

in the beginning there was experience
and perception of the qualities of experience
until we learned to label and name and categorize
the experiences
and the language of these labels and names and categories
began to frame our perceptions
to the point where perception is no longer in the body
but in the shadow of the body

> The very order that [labels and theories] provide engenders expectations that often impede fresh perception. Labels and theories provide a way of seeing. But a way of seeing is also a way of not seeing. (Eisner, 1998, p. 67)

Hillman (1982) states that in the ancient world, the heart was the organ of perception; it was the place of both sensation and imagination. He suggests that to "regard the world anew… we pay respect to it simply by looking again, re-specting, that second look with the eye of the heart" (p. 93). To look with the eye of the heart is to notice, before knowing; to perceive qualities; to not be limited by what we know and by what we can tell about what we know.

My participants spoke over and over again about their experiences of doing things over and over again, until these things were 'learned' or 'known' by heart.

"What seems crucial is the noticing, the active insertion of one's perception into the lived world... To ponder this is to become convinced that much of education as we know it is an education in forgetfulness. Distracting the young from their own perceived landscapes and shapes, we teachers insist on the givenness of predetermined explanatory frames." (Greene, 1995, p. 74)

I think it's because we did it over and over again and we just kept doing it. (Elsie, October 1999)

Throughout this text, I have criticized this meaningless approach to learning and knowing. The metaphorical critique of knowledge as a commodity for transmission, of the particle nature and shadowy texture of knowing, and of the hegemonic images of knowing (in) mathematics and science have enabled me to re-present the stories of my participants, embedded within a critical kaleidoscopic narrative, intended to both engage and provoke. The voices of my participants come through loud and (un)clear—they express much dissatisfaction with their school experiences of mathematics and science, but they remain stranded in given (and learned) frameworks for what counts as knowledge and what it means to know.

Sylvia & Harriet on "seeing" science...

Kathy: But science *does* come to you every day, doesn't it?
Harriet: Maybe it's because we don't SEE it *as* science.
Sylvia: Maybe it's because the questions asked aren't the same questions that *I* would ask or not the same way I would look at everything. So it always seems that I have to go to science. But maybe if different questions would be asked or things were approached in different ways, maybe science would come to me more. I don't know.
Harriet: I'd probably have a heart attack if it did. I'd be like 'whoa, something's really wrong with the world...'
Sylvia: I'd think 'ok, that's science but it's not the *right* science.' It's science but it's not the lab coat guy in glasses.
Harriet: I would think 'oh, that's too easy'. 'It's not the *real* science.' I would think 'it's wrong'.
Sylvia: 'That can't be real science'. Yeah, 'it makes *sense*, I get that... it can't be *real* science.'

Evelyn on "marks count"...

Kathy: ... about the teachers thinking you understand when you use formulas— I find that interesting to pursue. I certainly think my teachers were proud when I got 98% in grade ten math because I could do all the stuff. That made the teachers look like they were teaching well. And it made me look good. Everybody was happy. You know, the teacher's doing a good job because the students are getting good marks; the students are getting good marks so they're happy, they've got no complaints. So how can that ever change, do you think?

Evelyn: I think part of the problem there is that... the underlying idea that marks correspond to knowledge and for the most part—quite often— they don't. I think many teachers aren't evaluating their students on any kind of process of thought... they're looking for a product— it doesn't matter *how* you got there, or *why* you got there, just that you *get* there. And I think that's true— if you were going to be a teacher that stopped, let's say in grade 11 chemistry, and said 'well, we're going to do things differently', there would be a lot of students who would not be able to do it. They wouldn't be able to get away from the formulas and into the *why* and the *how* and that would cause problems because they haven't learned to think about science in a different kind of way. So, you'd have parents phoning. You'd have students not doing well. *And it would just be a mess*. Whereas, if you keep to how it's been, you'll have students who don't do well but you'll have students who can fill in the formulas, who can look at it and say: 'Well, I don't know how well I understand it, but it just works like this, and I don't know why.' And then they're happy because they get the 98 on the test and their parents are happy because it's 98 on the report card and the administration is happy because the students are doing reasonably well and they're going to get to go to university.

Kathy: What about for you— do you think your marks reflected understanding or knowledge?

Evelyn: Uh, uh, no. I think I probably did quite well in chemistry, but I don't understand chemistry at all. I don't really believe in atoms; I don't really know what they are—electrons, neutrons, protons. If I thought for a while, I could give you definitions for them. And if I thought a little bit longer, I could draw diagrams of how they fit together. But I don't know what they mean. But our test would ask 'what is an electron?' and you'd write down what it is and then it would either be right or wrong, but that doesn't demonstrate any kind of understanding. I can quite honestly say I don't understand chemistry... and even at a university level, I could do well in it because I could follow what they'd want me to do. I think that being a good student isn't so much about being particularly smart; it's about being able to figure out what your instructors want from you and being able to produce that. So, if you can do that, you can sail through quite a number of years, until you get somebody who asks you 'well, why is that?' and you think '*oh, why are you asking me this question* ? I want something that is a yes, or a no, that is factual; I don't want a *why* or a *how*... or a *compare*!'

Quote of the day...

They wouldn't be able to get away from the formulas and into the *why* and *how*
and that would cause problems because they haven't learned to think about science in a different kind of way.
So, you'd have parents phoning.
You'd have students not doing well.
And it would just be a mess.

(Evelyn, April 2000)

tread light(ly) on my knowledge

whose light travels in straight lines
reflecting off simple surfaces
all angles equal

when refraction is a choice?

why chip away at pieces of light
diffracted beyond recognition
through narrow pathways

when it needs only a change in wavelength?

how many interference patterns
must light run before reaching
its eventual dimness

when it can be constructive?

prove the laws of reflection
indices of refraction
critical angles
of light

but be
aware that I
have no place for more
or less useless knowledge

As McLaren (1994) points out, "knowledge must be made meaningful… before it can be made critical" (p. 224). Meaningless repetition of formula substitutions and definition memorization has led to an unfortunate interpretation of "knowing by heart".

know (verb)

know, savvy, ken, wot, wot of, ween
have knowledge, have a nodding acquaintance with, be acquainted
apprehend, conceive, catch, grasp, click, have, take in, get, understand
know entirely, possess, comprehend, master
come to know, latch on, get the hang of, get into one's head, realize
get to know, acquaint oneself, familiarize oneself, become au fait with
know again, recognize
know the value, appreciate
discern, discriminate
perceive, see

be in the know, be in the secret, have the lowdown, be informed
know by heart, know by rote, memorize
know backward and forward, have it down pat, have at one's finger tips, be master of,
know one's stuff, be expert
have a little knowledge of, not know
experience, know by experience, learn one's lesson, learn
get the picture, **see the light**
know all the answers, be omniscient
know what's what, see one's way, know one's way about, be wise.

(ORTEWP, 1994)

There are still the facts, the things you need to know by heart;
You just do them over and over again.
(Until your heart is just not in it anymore, if ever it was)

I called my sister the other day, just to reminisce and share school stories. Her 13-year old daughter answered the phone.

"Hey, whatcha doing?"
"Homework. Got a science test this week."

It was my opening. I could find out what it's all about to be a grade eight student learning science these days. After all, I've been told, classroom experiences of learning science have changed since I was a middle school student, and even since my participants were middle school students.

"What's your test on?" I quizzed.
"Chapter 3", she responded, confident that she need say no more.

Not being quite as enlightened as she imagined, I decided to explore a little further. Chapter 3, she informed me, is called "Face-lifting a Planet" and is, so far, all about volcanoes.

"What do you do for the test?"

Wondering if I am from another planet asking such a question, she explains carefully, "I have to study for it. I have a bunch of words—lava, erupt, magma, active, dormant, extinct—and I study their meanings. I also need to know the names and places of the..."

Her voice fades to soft whispers as I can hear her flipping through the pages of the textbook, counting.

"... ***18*** volcanoes in Canada."

Having momentarily caught a glimpse of a middle-school *deja vu*, I am determined to know more.

"So what kinds of things do you do in science class?"
"The teacher reads the textbook. She says 'you should know this and this...' and then she tells you what to copy from the text. We have to do a title page for every section."
"Oh, and is all your work copied from the science textbook?"
"No. Sometimes she'll put the words and meanings up on the overhead, and we copy them down. On the test, we give back what she gave us."

Intrigued by this simple portrayal of reciprocity in the classroom, I want to dig even deeper. I want to know how she *feels* about this. I wonder if I am climbing an active or dormant volcano.

"Do you like science?"
A pause. "Yeah. I like the experiments."
"Oh!" My heart skips a beat. "You do experiments?!"
"Yeah. We did one in chapter 1."
"Do you get to use your imagination?"

This one stumped her. (*Not a fair question, Aunty Kath.*)

"Ye..a..a..h... I guess. I got ***97*** in science."

When my sister came to the phone, I said: "Sounds like learning science is still all about knowing by heart."

She laughed. "Knowing by heart. I remember using that expression all the time. What does it *mean* anyway? Where did it *come* from?"

My sister continued, "School was always about memorizing and 'getting it into your head'. You don't stop to think 'am I learning?' You just want to give it back the way you got it."

The conversations made me ask 'what am **I** learning?'

I'm learning that my niece thinks I am a little out of touch with grade eight science, and perhaps she thinks I'm an extravagant aunt who asks peculiar questions about imagination in science class.

I'm learning that my sister is surprised by her daughter's 'young female' science teacher. "You'd think she'd be different", my sister pondered, at which point I shared my ideas on how I think it is difficult to be *different* when *same* feels more comfortable.

I'm learning that adherence to the dichotomous boundary between perceived playful science and the serious learning of science seems to be what middle school is all about. Aha! Now I think I know why it is called *middle* school—where playful learning moves through the in-between slash space to serious learning; where the meanings of the word **content** shift drastically from

> **con·tent** (kən-tĕnt′) *adjective*
> **1.** Desiring no more than what one has; satisfied.
> **2.** Ready to accept or acquiesce; willing: *She was content to step down after four years as chief executive.*

to

con·tent (kòn¹tènt´) *noun*

1. Often **contents .** Something contained, as in a receptacle: *the contents of my desk drawer; the contents of an aerosol can.*
2. Often **contents .** The subject matter of a written work, such as a book or magazine.

(AHDEL, 1992)

in such a manner that engaging enjoyment and imagination ('til your heart's content) is overshadowed by the belief that there are just so many facts you should know—things you just need to know by heart.

What possibilities can be imag(in)ed through this research text for a *different* "knowing by heart"?

To know by heart is a move into sensation *and* imagination because "the heart's way of perceiving is both a sensing and an imagining" (Hillman, 1982, p. 81).

> "But the role of imagination is not to resolve, not to point the way, not to improve. It is to awaken, to disclose the ordinarily unseen, unheard, and unexpected." (Greene, 1995, p. 28)

I understand different ways of teaching it and different ways of learning it— but still there's that answer at the end. Yeah, you can have fun and play with stuff to *get* to the answer but there's still the answer. (Helen, December 1999)

To know by heart is not pitting sensing *against* imagining, as "the game of oppositions" (Hillman, 1982, p. 92) would have us believe.

With science, they never asked you **'what do you think...?'** or **'how could you...?'** Science was always 'fill in this blank space here', right or wrong, half a mark here maybe, if you're lucky. (Helen, December 1999)

And I'm fine with the experiential: 'let's mess around with this'. That's good for me, but getting behind the scenes, to *teach* them the concepts, like why does this happen. There's always the concepts there... you *can't* go through the whole year just doing experiential learning in science—as much as I would *love* to do that—there's still underlying concepts that you have to learn, that are there and you just have to learn them; things you just have to know... (Sylvia, December 1999)

To know by heart is *poiesis*, a world in process, a text in the making.

I think the reality is there's so much out there— so much knowledge— that our human brains are just not able to retain it. (Roberta, January 2000)

They'll just learn what I'm able to teach them or learn on their own, but for me to be the knower of all, that's just impossible... makes me feel almost ashamed 'cause I won't be teaching them as well in science as I know I should be... but I don't know what to do about that. (Sylvia, December 1999)

... if *they're* learning something in science, chances are *I'm* going to have to learn it too so I can't SEE why they can't know that. (Ursula, October 1999)

To know by heart is to experience passion, commitment, and connection in all acts of thinking, questing, and knowing. This does not imply that knowing is unproblematic or unambiguous, but it also need not imply fear, or intimidation (or heart attack).

> ... the world is characterized as a process, a set of patterns, a web of dialectical interactions which the knower's passionate personal participation in events and the emotional, intuitive insight gained from such participation moves us toward a new dimension of cognition, a postformal mode of knowing. (Kincheloe et al., 1999, pp. 55-56)

I would say that I prefer to know things and to have people not know that I don't know things. I can put a hold on it and say 'ok well, I don't know right now but I can figure it out' and then I can probably find people or books that have the answer. So I can *do* those things. I don't think I'm missing a piece in my mind that doesn't permit me to *understand* science; I'm missing a lot of *desire* to know science... I think it's more my lack of *desire* that has stopped my growth—and lack of support and encouragement as well, I guess—

> rather than an inherent inability. (Evelyn, April 2000)
>
> In high school chemistry, I felt that I didn't even want to try anymore. I got to the point that I didn't even want to try anymore. I was like, 'what's the point? I'm not getting anywhere.' I remember feeling 'yuck, what's the point in even *coming* anymore?' I just felt discouraged. And that's how I left high school—just feeling discouraged with science. (Sylvia, October 1999)
>
> I don't think we feel *capable* of questioning [science]. I don't think it would ever cross my *mind* to question it because I feel like, well, I don't know enough to question it. (Ursula, October 1999)

To know by heart is to un/know what we think we do not know.

> I think [the experiment] challenges what you don't know. The fact that we couldn't get something to work and we couldn't tell you why it wasn't working, we just knew it was not working. We didn't know how to fix it. We couldn't find the answer. We tried. I think that's what's discouraging. You're like excited, excited, and then: 'you still know nothing.' For me, it was like, 'you still know nothing, so why are you bothering?' (Harriet, December 1999)
>
> And I think we're not *comfortable* with not knowing why. It's easier just to give up, then to try it another ten times and still have it not work. Sometimes it's just easier to give up and say 'ah, whatever, it's science!' ... It's that fear of unknowing. (Sylvia, December 1999)

The phenomenon of seeing is a very important consideration with/in the metaphor of light and knowing. In the words of Eisner (1998): "a way of seeing is also a way of not seeing" (p. 67). 'Seeing', as viewed in the light of this research text, is a personally transformative experience; an experience of "indwelling" (Polanyi, 1966). Seeing, like knowing, does not (or, at least, should not) consist of a predetermined framework, requiring everyone to 'see' and experience (this research text) in the same ways. Many people have responded to my research text in ways that I could not have imagined. I have been awakened to many interpretations of, and critical insights into, the possible meanings of my participants' lived experiences—from my participants *and* from the many RESPONSive readers.

To the reader...

Perhaps it is obvious to you by now that I am struggling with imagining possibilities for 'seeing' and knowing differently. I am taking my research questions to heart, so to speak, when I create these conc(ol)lusions with a determination not to ignore new spaces of
... questioning, without closing the search with answers
... interacting with ideas of knowing connected to one's own lived experience
... personal conscientization through narrative resonance

This aspect, I believe, is an empowering possibility for a critical narrative. My intention is not to impose my own filters such that the reader's engagement with this text is coloured entirely by my interpretations; it is not to limit the reader's vision by imposing my own myopic or hyperopic interpretations. What purpose would a kaleidoscope serve if I were to carefully and meticulously sketch out all of the patterns you ***should*** *have seen, instead of providing you with the opportunity to turn the kaleidoscope yourself. Such an approach would closely resemble my participants' experiences of mathematics and science—being told exactly what they are to know and how they should know it.*

As mentioned previously, I believe that *to know by heart* means we may have to learn to unknow what we think we do not know. Seeing is frequently limited by what we already (think we) know. I wonder, then, if one needs to unlearn and unknow in order to see anew.

How should I un/know?

It feels harsh to talk about a goal of
unknowing
so that I might begin again
and wonder
if in unlearning
the words and terms will
pain and stain
when they leave
now that I know the harm
engendered
when first they were consumed

Point me to the spaces
where unknowing occurs
my body still senses the pain of
ingesting the undigested
my mind still snaps and fidgets
then cedes
my heart is uneasy
it nearly takes me apart
a craving for I know not what
to live in such a space
to imagine more
yet imagine less
to see with the eyes of my heart

The Rule
for Ludwig Zeller

There is a rule
applicable to children –

which we should not
at any time forget.

Thus, when a bull's
discovered in the garden

or a unicorn
ascends the stairs,

we must not call the
wide-eyed tale a lie

or even speak about
imagination,

but accept realities;
we, too,

see what we do not see
and cannot know.

(Skelton, 1998, p. 214)

Reflexions...

I have invested a lot of time and energy into this research. This is not surprising. What does surprise me is the extent to which I have placed my heart into this research—my experiences, my truths; the shadows of my experiences and my truths. I write of the importance of preservice teachers' reflections on their lived experiences of learning and knowing (in) mathematics and science... but what happens if it takes them on a similar journey to my own? What happens if someone, who used to love teaching, becomes so disillusioned with it all that she does not know how to begin to imagine things differently but knows, in her heart, that she must?

In 'seeing' this storied research text as a personal journey, there are still reasons why the stories have been written. Code (1995) asks the question:

"But what, at the end of this story, is knowledge?" (p. 184)

Knowledge is engaging with this text to catch a glimpse of how little we engage with the world. Hillman (1982) writes of the importance of appreciation and noticing before we can know: "Perhaps, events speed up in proportion to their not being appreciated; perhaps events grow to cataclysmic size and intensity in proportion to their not being noticed" (p. 85).

Knowledge is acknowledging the stories of each of these preservice teachers' experiences. It is about asking the critical question of how we can continue to feign surprise and disbelief when we, as educators, hear the stories of frustrating and meaningless experiences in/of school mathematics and science.

Knowledge is taking these stories to heart and responding to the transformative possibilities that exist in localized epistemological interrogations. Such interrogations might include the politics of schooling, the cultural landscapes of mathematics and science, and the hegemony of curricular knowledge.

Knowledge is respecting the kaleidoscopic image metaphor of lived experience narratives, or "little stories", so that grand narratives may be seen as (un)easy generalizations and escapist ideologies.

> Counternarratives... counter not merely (or even necessarily) the grand narratives, but also (or instead) the "official" and "hegemonic" narratives of everyday life. (Peters & Lankshear, 1996, p. 2)
>
> ...I find the very effort to shape the materials of lived experience into narrative to be a source of meaning making. It is because we are reflecting back when we tell our stories, that we may be able to recapture the nascent *logos*—our mind just before it emerged from the perceived and vivid and began abstracting. (Greene, 1995, p. 75)

Knowledge is living and telling the critical and partial stories that shape us as teachers; the stories that confine and restrict, that limit and delineate, that once and for all move us to spaces of dissatisfaction and dissonance; the stories that compel us to deconstruct the hegemonic "truths" of schools' purposes and goals.

A final focus group reflection on knowing (in) math and science...

Evelyn: One of the things I was thinking... we don't ever really think that there's room to ***be*** creative in math; that there ***is*** only one way. And if ***you*** think only one way about math, how do you encourage that kind of thought with your students?

Kathy: I was talking with one of the participants and it was interesting... instead of looking at how much ***you know*** about math and science, what about looking at math and science themselves and how they're portrayed to you. One of the conversation comments was: "well, maybe math and science ***could*** be different but ***I've*** never met a math or science ***yet*** that I like." I thought that was an interesting way of putting it. Maybe there ***is*** a way of doing math that's creative; maybe there is a way of doing science that's creative and more holistic, instead of picking on the grammar of science... but did any of us ever experience that kind of thing?

Harriet: I know that now we've got the coolest stuff in math— manipulatives and all these games for every level and adaptive dimension. But I think that the fear is... maybe ***my*** fear is, you get ***so used*** to how you were taught— and I don't want to teach it like that because I think that having *total* seatwork and *total* bookwork... oh, that's not teaching kids ***anything***. I mean, it will work for some but for those who need to touch things, to be involved... it's sometimes hard for us to come into this new place with centres and blocks and... it's hard because you're still stuck at the point where you were at and then [we're expected] to go to this different level of teaching. I think that's kind of hard. I would be afraid to go too far one way with *all* manipulatives and hands-on and then the solid, concrete things *aren't* being taught. Do you know what I mean? I think we can go the spectrum, one way or another... I'm worried about not finding that happy medium where there are manipulatives but yet there is also ***solid work***, like multiplication or whatever, things that children ***do*** need to know. Like, where's that happy medium? Because I don't think you always find that.

Kathy: It's the question of what's important knowledge... for so many years you get a lot of repeated questions... that was something that came up in our conversations; that if you introduce a whole new kind of curriculum or textbook, sometimes teachers don't want to use them because the students won't be getting ***'the math they need to know'***... the idea that there **is** a math you need to know that's been handed down; a science you need to know that's been handed down. But whoever decided that that's really a basic skill you need in life, or is it something we go with because of tradition?

Roberta: I think the part about the transition from elementary school to high school is also an important point. I know grade eight teachers, looking at their class and saying ***'they need to know this, this and this before they get to high school'***... if they don't ***have*** those skills, they're not going to get them in high school because [the skills] are not going to be taught. And I think, as elementary teachers, we have to be able to ***give*** the skills, so that when the kids ***do*** hit high school they do have the skills to progress through those next four years and whatever career choices they plan… that they have the skills; that they're rooted in elementary.

Kathy: We've all experienced those four years and we all have memories…

Sylvia: I think we talked about this before but even with let's say, self-confidence in girls in math and science… we do all these nifty things as elementary teachers to build up ***every*** child's confidence and self-esteem so ***they can do the subject*** and then all of a sudden they hit high school and then I don't know what happens… 'cause I remember, I have *awful* memories from math and science in high school and so now I'm wondering— right now we're doing all these lovely things to get the kids built up so they have high confidence and motivation but then, somehow, something happens in high school. I don't know.

APPENDIX

"Answers in the Back"

CHAPTER ONE
Where does light come from? REVIEW

Criss-cross puzzle words:

Across
2. livingroom
4. knowlous
6. metonomy
8. western
9. enlightenment
10. discursive
11. doubling
12. presence

Down
1. dichotomy
2. luminous
3. objective
5. objectivity
7. expertise

CHAPTER TWO
How does light travel? REVIEW

Word Search with Hidden Message

N	O	I	T	C	A	R	F	E	R	T	H	G	I	L	T	T
T	T	I	N	L	A	N	G	U	A	G	E	M	E	E	N	E
A	R	E	C	T	I	L	I	N	E	A	R	T	X	O	E	X
C	R	E	I	D	E	O	L	O	G	Y	F	T	L	E	R	T
I	C	T	H	Y	O	R	E	L	C	I	T	R	A	P	A	B
T	N	Y	D	O	O	U	F	E	C	L	I	P	S	E	P	O
T	R	A	N	S	L	U	C	E	N	T	R	T	E	P	S	O
N	H	N	A	C	T	O	H	I	R	N	G	A	N	R	N	K
S	O	E	O	D	L	L	G	E	A	E	R	N	I	O	A	F
M	S	I	V	I	T	C	U	R	T	S	N	O	C	P	R	L
N	G	+	T	A	T	+	N	A	A	+	+	C	+	A	T	A
+	+	+	+	P	W	C	+	O	C	P	+	+	E	G	U	T
+	+	S	E	I	R	O	E	H	T	U	H	+	U	A	M	L
W	O	D	A	H	S	O	+	L	+	O	O	Y	Q	T	B	A
G	N	I	W	O	N	K	S	+	F	+	H	F	A	I	R	N
A	R	B	M	U	N	E	P	B	+	E	+	P	P	O	A	D
H	E	G	E	M	O	N	Y	+	A	+	R	+	O	N	+	+

(Over, Down, Direction)
ABSORPTION (10,17,NW)
CONSTRUCTIVISM (14,10,W)
ECLIPSE (9,6,E)
FLATLAND (17,9,S)
FOUCAULT (13,15,NW)
HEGEMONY (1,17,E)
HOLOGRAPHY (4,5,SE)

IDEOLOGY (4,4,E)
INTERFERENCE (3,1,SE)
KNOWING (7,15,W)
LANGUAGE (5,2,E)
LIGHT (15,1,W)
OPAQUE (14,17,N)
PARTICLE(15,5,W)
PENUMBRA (8,16,W)
PHOTON (13,16,NW)
PROPAGATION (15,7,S)
RECTILINEAR (2,3,E)
REFLECTION (12,17,NW)
REFRACTION (10,1,W)
SHADOW (6,14,W)
SHADY (1,9,NE)
TACIT (1,2,S)
TEXT (16,1,SW)
TEXTBOOK (17,1,S)
THEORIES (10,13,W)
RANSLUCENT (1,7,E)
TRANSPARENT (16,11,N)
UMBRA (16,12,S)
WAVE (6,12,NW)

Hidden Message: **TIME TO REFLECT ON YOUR TEACHING AND LEARNING**

CHAPTER THREE
What happens when light hits different surfaces? REVIEW

The **clue words** are:

Absorption
Reflection
Refraction
Light
Ignorance
Silence
Resistance
Hegemony
Colour
Culture
Meaningless

Hidden phrase: **REFLECT ON REFRACTION**

CHAPTER FOUR
How are images formed? REVIEW

Fallen Phrase: **SCIENCE REIMAGINES ITSELF IN DIFFERENT LENSES OF KNOWING**

NOTES

[1] The drawings of the atoms were completed by preservice teachers in elementary science education classes, 1998-1999

[2] Background constellation image reference: Levy (1994)

[3] Before delving any further into this text, I want to shed some light on a playful feature of perFORMing the text—that is, the use of parentheses, slashes, and mixed cased letters within words and phrases; I have employed such textual 'play' in an attempt to create openings for the reader—to encourage multiple ways of reading and interpreting the words and/or phrases. For example, are you able to imagine different meanings associated with knowing mathematics and knowing **in** mathematics?

[4] Many thanks to Dr. Ted Aoki and Dr. Marilyn Low, Summer Institute, University of Lethbridge, July 12, 2000

[5] End of the end-of-chapter review puzzles were created using Discovery School's Puzzlemaker; Available: http://www.puzzlemaker.com

REFERENCES

Abbott, E.A. (1984). *Flatland: A romance of many dimensions.* Ontario: The New American Library of Canada Limited.

AHDEL (*The American Heritage® Dictionary of the English Language, Third Edition*) (1992). Electronic version licensed from INSO Corporation. United States: Houghton Mifflin Company.

Alrø, H. & Skovsmose, O. (2002). *Dialogue and learning in mathematics education: Intention, reflection, critique.* New York, NY: Kluwer.

Aoki, T. (1999). Interview with Ted Aoki: Rethinking curriculum and pedagogy. *Kappa Delta Pi Record, Summer 1999*, 180-181.

Aoki, T. (2000). *Locating living pedagogy in teacher "research": Five Metonymic moments.* Paper presented at the "International Conference on Teacher Research", Baton Rouge, LA. April, 2000.

Aoki, T. (2003). Locating living pedagogy in teacher "research": Five metonymic moments. In E. Hasbe-Ludt & W. Hurren (Eds.), *Curriculum intertext: Place/language/pedagogy* (pp. 1-9). New York: Peter Lang Publishing, Inc.

Apple, M.W. (1986). *Teachers and texts: A political economy of class and gender relations in education.* New York: Routledge and Kegan Paul.

Apple, M.W. (1990). *Ideology and curriculum* (second edition). New York: Routledge.

Apple, M.W. (1993). *Official knowledge: Democratic education in a conservative age.* New York: Routledge.

Apple, M.W. (1995). *Education and power* (second edition). New York: Routledge.

Apple, M. (2004). *Ideology and curriculum (third edition).* New York: RoutledgeFalmer.

Ausubel, D. (1968). *Educational Psychology.* New York: Holt, Reinhart, & Winston.

Ball, D. (1990). Breaking with experience in learning to teach mathematics: the role of a preservice methods course. *For the Learning of Mathematics, 10,* 10-16.

Ball, D. & Bass, H. (2000). Interweaving content and pedagogy in teaching and learning to teach: knowing and using mathematics. In J. Boaler (Ed.), *Multiple perspectives on mathematics teaching and learning* (pp. 83-104). Westport, CT: Greenwood Publishing Group.

Barton, A.C. (1998). *Feminist science education.* New York: Teachers College Press.

Barton, A.C. (2003). *Teaching science for social justice.* New York: Teachers College Press.

Bauer, H. H. (1992). Scientific literacy and the myth of the scientific method. Chicago: University of Illinois Press.

Beeth, M. (1998). Facilitating conceptual change learning: The need for teachers to support metacognition. *Journal of Science Teacher Education, 9*, 49-61.

Belenky, M.F. (1996). Public homeplaces: Nurturing the development of people, families, and communities. In Goldberger, Tarule, Clinchy, & Belenky (Eds.), *Knowledge, difference, and power: Essays inspired by Women's Ways of Knowing* (pp. 393-430). New York: Basic Books.

Belenky, M.F., Clinchy, B.M., Goldberger, N.R., & Tarule, J.M. (1986). *Women's ways of knowing: The development of self, voice, and mind.* New York: Basic Books, Inc.

Bernstein, R. (1976). *The restructuring of social and political theory.* United States: University of Pennsylvania Press.

Bernstein, R.J. (1991). *The new constellation: The ethical-political horizons of modernity/postmodernity.* Cambridge, Mass.: The MIT Press.

Bozin-Mirkovic, I. (1997). Preservice teachers' voices: Images of teaching and the self. Unpublished doctoral dissertation, University of Massachusetts, Amherst.

Briscoe, C., & Stout, D. (1996). Integrating math and science through problem centered learning in methods courses: Effects on prospective teachers' understanding of problem solving. *Journal of Elementary Science Education, 8*, 66-87.

Britzman, D. (1990). The terrible problem of knowing thyself: Toward a poststructural account of teacher identity. *Journal of Curriculum Theorizing, 9*(3), 23-46.

Britzman, D.P. (1991). *Practice makes practice: A critical study of learning to teach.* New York: State University of New York Press.

Britzman, D. (2003). *Practice makes practice: A critical study of learning to teach (revised edition).* New York: State University of New York Press.

Bruenn, S.W. (1968). *College board achievement test physics.* New York: Arc Books, Inc.

Bruner, J.S. (1979). *On knowing: Essays for the left hand (expanded edition).* Cambridge, MA.: The Belknap Press of Harvard University Press.

Bruner, J.S. (1996). *The culture of education.* Cambridge, MA.: Harvard University Press.

Bulla, C.R. (1994). *What makes a shadow?* (revised edition). New York: HarperCollins Publishers.

Burbules, N.C. & Berk, R. (1999). Critical thinking and critical pedagogy: Relations, differences, and limits. In T.S. Popkewitz & L. Fendler (Eds.), *Critical theories in education: Changing terrains of knowledge and politics* (pp. 45-66). New York: Routledge.

Burton, L. (1996). Mathematics, and its learning, as narrative— A literacy for the twenty-first century. In D. Baker, J. Clay, & C. Fox (Eds.), *Challenging ways of knowing: In English, Maths and Science* (pp. 13-18). London: Falmer Press.

Carspecken, P.F. & Apple, M. (1992). Critical qualitative research: Theory, methodology, and practice. In M.D. LeCompte, W.L. Millroy, & J. Preissle (Eds.), *The handbook of qualitative research in education* (pp. 507-553). San Diego: Academic Press, Inc.

Cato, D. (2006). Of the trivial and the radical: Is there a coherent constructivist pedagogy? *Paideusis, 15*(1), 57-74.

Cixous, H. & Calle-Gruber, M. (1997). *Hélène Cixous Rootprints: Memory and life writing.* New York: Routledge.

Clandinin, D.J. & Connelly, F.M. (2000). *Narrative inquiry: Experience and story in qualitative research.* California: Jossey-Bass Inc., Publishers.

Clay, J. (1996). Scientific literacy: Whose science? Whose literacy? In D. Baker, J. Clay, & C. Fox (Eds.), *Challenging ways of knowing: In English, Maths and Science* (pp. 184-193). London: Falmer Press.

Code, L. (1991). *What can she know? Feminist theory and the construction of knowledge.* New York: Cornell University Press.

Code, L. (1995). *Rhetorical spaces: Essays on gendered locations.* New York: Routledge.

Cohen, L. (1998). Anthem. In S. McCaslin (Ed.), *A matter of spirit: Recovery of the sacred in contemporary Canadian poetry.* Victoria, BC: Ekstasis Editions Canada Ltd.

Collins, J. & Mayblin, B. (1996). *Derrida for beginners.* Great Britain: Icon Books Ltd.

Conle, C. (1996). Resonance in preservice teacher inquiry. *American Educational Research Journal, 33*(2), 297-325.

Cooney, T.J., Shealy, B.E., & Arvold, B. (1998). Conceptualizing belief structures of preservice secondary mathematics teachers. *Journal for Research in Mathematics Education, 29,* 306-333.

Danielewicz, D. (2001). *Teaching selves: Identity, pedagogy, and teacher education.* New York: State University of New York Press.

Daughter can count on the future. (2000, September 19). *Leader Post,* p. C1.

Davies, B. (2000). *A body of writing, 1990-1999.* New York: Alta Mira Press.

Davis, A.B., Sumara, D.J., & Kieren, T.E. (1996). Cognition, co-emergence, curriculum. *Journal of Curriculum Studies, 28*(2), 151-169.

Davis, F. & Steiger, A. (1993). Feminist pedagogy in the physical sciences. ERIC Document ED372781.

Debold, E., Tolmam, D., & Brown, L.M. (1996). Embodying knowledge, knowing desire: Authority and split subjectivities in girls' epistemological development. In Goldberger, Tarule, Clinchy, & Belenky (Eds.), *Knowledge, difference, and power: Essays inspired by Women's Ways of Knowing* (pp. 85-116). New York: Basic Books.

de Freitas, E. & Nolan, K. (in press). *Opening the research text : Critical insights and in(ter)ventions into mathematics education.* New York: Springer.

Derrida, J. (1972). *Dissemination.* [Translated, with an introduction and additional notes by B. Johnson (1982).] Chicago: The University of Chicago Press.

Dewey, J. (1938). *Experience and education.* New York: Kappa Delta Pi.

Dimitriadis, G. & Kamberelis, G. (2006). *Theory for education.* New York: Routledge.

Dorey, A.A. (1949). *Light and Shadow*. Toronto: The Ryerson Press.

Driver, R., & Erickson, G. (1983). Theories-in-action: Some theoretical and empirical issues in the study of students' conceptual frameworks in science. *Studies in Science Education, 10*, 37-60.

Duit, R. (1995). The constructivist view: A fashionable and fruitful paradigm for science education and practice. In L.P. Steffe & J. Gale (Ed.), *Constructivism in education* (pp. 271-285). New Jersey: Lawrence Erlbaum Associates, Inc.

Duit, R. & Treagust, D.F. (2003). Conceptual change: a powerful framework for improving science teaching and learning. *International Journal of Science Education, 25*(6), 671-688.

Eisner, E.W. (1998). *The enlightened eye: Qualitative inquiry and the enhancement of educational practice*. New Jersey: Prentice-Hall, Inc.

Ernest, P. (2004). Postmodernism and the subject of mathematics. In M. Walshaw (Ed.), *Mathematics education within the postmodern* (pp. 15-33). Greenwich, CN: Information Age Publishing

Fearing, S. (2000). Dog on a chain. On *So many miles* [CD]. Toronto: True North.

Fels, L. and Meyer, K. (1997). On the edge of chaos: Co-evolving world(s) of drama and science. *Teaching Education, 9*(1), 75-81.

Foucault, M. (1990). *The history of sexuality: An introduction (volume1)*. New York: Random House, Inc.

Franklin, U. (1995). Looking forward, looking back. Paper presented at "More Than Just Numbers" Conference on Women and Engineering at the University of New Brunswick, May 1995. Retrieved August 5, 2000 from the World Wide Web: http://www.coolwomen.org/coolwomen/

Gelwick, R. (1977). *The way of discovery: An introduction to the thought of Michael Polanyi*. New York: Oxford University Press.

Giroux, H.A. (1981). *Ideology, culture, and the process of schooling*. Philadelphia: Temple University Press.

Giroux, H.A. (1983). *Theory and resistance in education: A pedagogy for the opposition*. Massachusetts: Bergin & Garvey Publishers, Inc.

Giroux, H.A. (1989). Schooling as a form of cultural politics: Toward a pedagogy of and for difference. In H.A. Giroux & P.L. McLaren (Eds.), *Critical pedagogy, the state, and cultural struggle*. New York: State University of New York Press.

Giroux, H.A., Lankshear, C., McLaren, P. & Peters, M. (Eds.) (1996). *Counternarratives: Cultural studies and critical pedagogies in postmodern spaces*. New York: Routledge.

Goldberger, N.R. (1996a). Looking backward, looking forward. In Goldberger, Tarule, Clinchy, & Belenky (Eds.), *Knowledge, difference, and power: Essays inspired by Women's Ways of Knowing* (pp. 1-21). New York: Basic Books.

Goldberger, N.R. (1996b). Cultural imperatives and diversity in ways of knowing. In Goldberger, Tarule, Clinchy, & Belenky (Eds.), *Knowledge, difference, and power: Essays inspired by Women's Ways of Knowing* (pp. 335- 371). New York: Basic Books.

Golden, K.L. (Ed.) (1992). *Uses of comparative mythology: Essays on the work of Joseph Campbell*. New York: Garland Publishing, Inc.

Goodnough, K. & Cashion, M. (2003). Fostering inquiry through problem-based learning. *The Science Teacher*, December, 21-25.

Goos, Galbraith, and Renshaw (2004). Establishing a community of practice in a secondary mathematics classroom. In B. Allen & S. Johnston-Wilder (Eds.), *Mathematics education: Exploring the culture of learning* (pp. 91-116). London: RoutledgeFalmer.

Gordon, M. (2006). Welcoming confusion, embracing uncertainty: Educating teacher candidates in an age of certitude. *Paideusis, 15*(2), 15-25.

Greene, M. (1973). *Teacher as stranger: Educational philosophy for the modern age*. California: Wadsworth Publishing Company.

Greene, M. (1995). *Releasing the imagination: Essays on education, the arts, and social change*. California: Jossey-Bass Inc., Publishers.

Griffin, S. (1995). *The Eros of everyday life*. New York: Doubleday.

Griffiths, M. (2006). The feminization of teaching and the practice of teaching: Threat or opportunity? *Educational Theory, 56*(4), 387- 405.

Grow, P. & Plucker, J. (2003). Good problems to have: Implementing problem-based learning without redesigning a curriculum. *The Science Teacher*, December, 31-35.

Gutstein, E. (2006). *Reading and writing the world with mathematics: Toward a pedagogy for social justice*. New York: Routledge.

Habermas, J. (1971). *Knowledge and human interests.* Boston, MA: Beacon Press.

Harding, S. (1992). The instability of the analytical categories of feminist theory. In H. Crowley & S. Himmelweit (Eds.), *Knowing women: Feminism and knowledge* (pp. 338-354). Cambridge, UK: The Open University.

Hashweh, M.Z. (1996). Effects of science teachers' epistemological beliefs in teaching. *Journal of Research in Science Teaching, 33*, 47-63.

Hecht, J. (1987). *Optics: Light for a new age*. New York: Charles Scribner's Sons.

Hekman, S.J. (1990). *Gender and knowledge: Elements of a postmodern feminism.* Boston: Northeastern University Press.

Hewson, P., & Thorley, R. N. (1989). The conditions of conceptual change in the classroom. *International Journal of Science Education, 11*, 541-553.

Hildebrand, G.M. (1998). Disrupting hegemonic writing in school science: Contesting the right way to write. *Journal of Research in Science Teaching, 35*, 345-362.

Hill, L. (1997). Just tell us the rule: Learning to teach elementary mathematics. *Journal of Teacher Education, 48*, 211-221.

Hillman, J. (1982). Anima Mundi: The return of the soul to the world. *Spring 1982*, 71-93.

Horus Publications on the Internet (August 2000). *Michel Foucault.* Available: [http://www.lucknow.com/horus/guide/cm108.html].

Hurtado, A. (1996). Strategic suspensions: Feminists of color theorize the production of knowledge. In Goldberger, Tarule, Clinchy, & Belenky (Eds.), *Knowledge, difference, and power: Essays inspired by Women's Ways of Knowing* (pp. 372-392). New York: Basic Books.

Hutchinson (1995). *The Hutchinson concise science dictionary.* Oxford: Helicon Publishing Ltd.

Janesick, V.J. (1998). The dance of qualitative research design: Metaphor, methodolatry, and meaning. In N.K. Denzin & Y.S. Lincoln (Eds.), *Strategies of qualitative inquiry* (pp. 35-55). California: Sage Publications.

Jenkins, F.A. & White, H.E. (1976). *Fundamentals of optics (fourth edition).* New York: McGraw-Hill, Inc.

Johansson, M. (2007). Mathematical meaning making and textbook tasks. *For the learning of mathematics, 27*(1), 45-51.

Jones, E.R. & Childers, R.L. (1993). *Contemporary college physics (second edition).* Reading, MA: Addison-Wesley Publishing Company.

Kajander, A. & Dionne, J. (2005). Elementary teacher education: Defining the crucial experiences. In E. Simmt & B. Davis (Eds.), *Proceedings of the 2004 Annual Meeting of the Canadian Mathematics Education Study Group (CMESG).* Edmonton, AB: CMESG.

Kang, H. (1996). Preservice teachers' beliefs and conceptions of mathematics learning and teaching. Unpublished master's thesis, University of Regina, Saskatchewan.

Keller, E.F. (1995). *Reflections on gender and science (10th Anniversary Edition).* New Haven: Yale University Press.

Kelsey, M.T. (1991). *God, dreams, and revelation: A Christian interpretation of dreams.* Minneapolis, MN: Augsburg.

Keys' C. & Bryan, L. (2001). Co-constructing inquiry-based science with teachers: Essential research for lasting reform. *Journal of Research in Science Teaching, 38*(6), 631–645.

Kincheloe, J.L., Steinberg, S.R., & Tippins, D.J. (1999). *The stigma of genius: Einstein, consciousness, and education.* New York: Peter Lang Publishing, Inc.

Kuhn, T.S. (1996). *The structure of scientific revolutions (third edition).* Chicago: The University of Chicago Press.

Kurth, L.A. (1995). Examining the significance of science in the lives of prospective elementary teachers. ERIC Document ED 391 643

Lather, P. (1991). *Getting smart: Feminist research and pedagogy with/in the postmodern.* New York: Routledge, Chapman and Hall, Inc.

Lather, P. (1999). From competing paradigms to disjunctive affirmation. Paper presented at the Annual Conference of the American Educational Research Association, Montreal, Quebec, April 1999.

LeCourt, D. (2004). *Identity matters: Schooling the student body in academic discourse.* New York, NY: State University of New York Press.

Lee, D. (1995). Poetry and unknowing. In T.Lilburn (Ed.), *Poetry and knowing* (pp. 29-44). Ontario: Quarry Press, Inc.

Leggo, C. (1999). Research as poetic rumination: Twenty-six ways of listening to light. *Journal of Educational Thought, 33*(2), 113-33.

Lemke, J.L. (1990). *Talking science: Language, learning, and values.* New Jersey: Ablex Publishing Corporation.

Levy, D.H. (1994). *Skywatching: The nature companion guides.* Australia: Weldon Owen Inc.

Lewis, M.G. (1993). *Without a word: Teaching beyond women's silence.* New York: Routledge.

Lightfoot, S. (1999). In *The language of teaching: Thoughts on the art of teaching and the meaning of education.* Colorado: Blue Mountain Press.

Lincoln, Y.S. (1997). Self, subject, audience, text: Living at the edge, writing in the margins. In W.G. Tierney & Y.S. Lincoln (Eds.), *Representation and the text: Re-framing the narrative voice* (pp. 37-56). New York: State University of New York Press.

Maxwell, J.A. (1996). *Qualitative research design: An interactive approach.* California: Sage Publications, Inc.

McLachlan, S. (1993). Good enough. On *Fumbling towards ecstasy* [CD]. Vancouver: Nettwerk Productions.

McLaren, P. (1994). *Life in schools: An introduction to critical pedagogy in the foundations of education (2nd edition).* New York: Longman.

Meyer, K. (1998). Reflections on being female in school science: Toward a praxis of teaching science. *Journal of Research in Science Teaching, 35*(4), 463-471.

Michael, M. (1996). Ignoring science: Discourses of ignorance in the public understanding of science. In A. Irwin & B. Wynne (Eds.), *Misunderstanding science? The public reconstruction of science and technology* (pp. 107-125). Great Britain: Cambridge University Press.

Michaels, A. (1995). Cleopatra's Love. In T.Lilburn (Ed.), *Poetry and knowing* (pp. 177-183). Ontario: Quarry Press, Inc.

Millay, E.S.V. (1941). *Collected Sonnets of Edna St. Vincent Millay.* New York: Harper & Brothers Publishers.

Morgan, C., Watson, A. & Tikly, C. (2004). *Teaching school subjects 11-19: Mathematics.* London, UK: RoutledgeFalmer.

Mulholland, J., & Wallace, J. (1996). Breaking the cycle: Preparing elementary teachers to teach science. *Journal of Elementary Science Education, 8*, 17-38.

National Council of Teachers of Mathematics (NCTM) (2003). *Integrated mathematics: Choices and challenges.* Reston, VA: NCTM, Inc.

Noddings, N. & Shore, P.J. (1984). *Awakening the inner eye: Intuition in education.* New York: Teachers College Press.

Nolan, K. (2005). Publish or cherish? Performing a dissertation in/between research spaces. In R. Barnett (Ed.), *Reshaping the university: New relationships between research, scholarship and teaching* (pp. 119-135). Great Britain: Open University Press.

O'Connor, P. (1986). *Dreams and the search for meaning.* New York: Paulist Press.

ORTEWP *(The Original Roget's Thesaurus of English Words and Phrases* (Americanized Version) (1994). Electronic version licensed from Longman Group. United Kingdom: Longman Group UK Limited.

Peters, M. & Lankshear, C. (1996). Postmodern counternarratives. In H.A. Giroux, C. Lankshear, P. Mclaren, & M. Peters, *Counternarratives: Cultural studies and critical pedagogies in postmodern spaces* (pp. 1-39). New York: Routledge.
Polanyi, M. (1966). *The tacit dimension*. Great Britain: Routledge & Kegan Paul Ltd.
Pope, M., & Gilbert, J. (1983). Personal experience and the construction of knowledge in science. *Science Education, 67*, 193-203.
Popkewitz, T.S. (1999). Critical traditions, modernisms, and the "posts". In T.S. Popkewitz & L. Fendler (Eds.), *Critical theories in education: Changing terrains of knowledge and politics* (pp. 1-13). New York: Routledge.
Posner, G., Strike, K., Hewson, P., & Gertzog, W. (1982). Accommodation of a scientific conception: Toward a theory of conceptual change. *Science Education, 66*, 211-227.
Povey, H. & Burton, L. (2004). Learners as authors in the mathematics classroom. In B. Allen & S. Johnston-Wilder (Eds.), *Mathematics education: Exploring the culture of learning* (pp. 43-56). London: RoutledgeFalmer.
Richardson, L. (1992). The poetic representation of lives: Writing a postmodern sociology. *Studies in Symbolic Interaction, 13*, 19-27.
Richardson, L. (1998). Writing: A method of inquiry. In N. Denzin & Y. Lincoln (Eds.), *Collecting and interpreting qualitative materials* (pp. 345-371). California: Sage Publications, Inc.
Riggs, I. M. (1991). Gender differences in elementary science teacher self-efficacy. ERIC Document ED 340 705
Rockhill, K. & Tomic, P. (1995). Situating ESL between speech and silence. In J. Gaskell & J.Willinsky (Eds.), *Gender in/forms curriculum: From enrichment to transformation*. New York: Teachers College Press.
Rose, H. (1994). *Love, power and knowledge: Towards a feminist transformation of the sciences*. Cambridge, UK: Polity Press.
Rosser, S.V. (1997). *Re-engineering female friendly science*. New York: Teachers College Press.
Rossman, G.B. & Rallis, S.F. (1998). *Learning in the field: An introduction to qualitative research*. California: Sage Publications, Inc.
Roth, W.-M. (1995). *Authentic school science: Knowing and learning in open-inquiry science laboratories*. The Netherlands: Kluwer Academic Publishers.
Roth, W.-M. (Ed.) (2005). *Auto/biography and auto/ethnography: Praxis of research method*. The Netherlands: Sense Publishers.
Roth, W.-M. & Barton, A.C. (2004). *Rethinking scientific literacy*. New York: RoutledgeFalmer.
Roth, W.-M. & Desautels, J. (Eds.) (2002). *Science education as/for sociopolitical action*. New York: Peter Lang.
Roychoudhury, A., Tippins, D.J., & Nichols, S.E. (1993). An exploratory attempt toward a feminist pedagogy for science education. *Action in Teacher Education, 25*, 36-46.
Roychoudhury, A., Tippins, D.J., & Nichols, S.E. (1995). Gender-inclusive teaching: a feminist – constructivist approach. *Journal of Research in Science Teaching, 32*, 897-924.
Ruddick, S. (1996). Reason's feminity: A case for connected knowing. In Goldberger, Tarule, Clinchy, & Belenky (Eds.), *Knowledge, difference, and power: Essays inspired by Women's Ways of Knowing* (pp. 248-273). New York: Basic Books.
Schoon, K.J. (1995). The origin and extent of alternative conceptions in the earth and space sciences: A survey of pre-service elementary teachers. *Journal of Elementary Science Education, 7*, 27-46
Scott, J. (1997). Deconstructing equality-versus-difference: Or, the uses of poststructuralist theory for feminism. In D.T. Meyers (Ed.), *Feminist social thought* (pp. 758-770). New York: Routledge.
Skelton, R. (1998). Knowledge. In S. McCaslin (Ed.), *A matter of spirit: Recovery of the sacred in contemporary Canadian poetry* (pp. 214-215). Victoria, BC: Ekstasis Editions Canada Ltd.
Skelton, R. (1998). The Rule. In S. McCaslin (Ed.), *A matter of spirit: Recovery of the sacred in contemporary Canadian poetry* (p. 214). Victoria, BC: Ekstasis Editions Canada Ltd.
Skovsmose, O. (2005). *Travelling through education: Uncertainty, mathematics, responsibility*. The Netherlands: Sense Publishers.

Smith, D. & Neale, D. (1989). The construction of subject matter knowledge in primary science teaching. *Teaching and Teacher Education, 5*, 1-20.

Smith, D.E. (1991). An analysis of ideological structures and how women are excluded: Considerations for academic women. In J.S. Gaskell & A.T. McLaren (Eds.), *Women and education (2nd edition)* (pp. 233-256). Calgary, Alberta: Detselig Enterprises Limited.

Stevens, R. (2000). Who counts what as math? Emergent and assigned mathematics problems in a project-based classroom. In J. Boaler (Ed.), *Multiple perspectives on mathematics teaching and learning* (pp. 105-144). Westport, CT: Ablex Publishing.

Stofflett, R., & Stefanon, L. (1996). Elementary teacher candidates' conceptions of successful conceptual change teaching. *Journal of Elementary Science Education, 8*, 1-20.

Stofflett, R.. & Stoddart, T. (1994). The ability to understand and use conceptual change pedagogy as a function of prior content learning experience. *Journal of Research in Science Teaching, 31*, 31-51.

Sutton, C. (1996). Beliefs about science and beliefs about language. *International Journal of Science Education, 18*(1), 1-18.

Threadgold, T. (1997). *Feminist poetics: Poiesis, performance, histories*. London: Routledge.

Tierney, W.G. (1997). Lost in translation: Time and voice in qualitative research. In W.G. Tierney & Y.S. Lincoln (Eds.), *Representation and the text: Re-framing the narrative voice* (pp. 23-36). New York: State University of New York Press.

Tillema, H.H., & Knol, W.E. (1997). Promoting student teacher learning through conceptual change or direct instruction. *Teaching and Teacher Education, 13*, 579-595.

Tobin, K., Tippins, D., & Gallard, A. (1994). Research on instructional strategies for teaching science. In D. Gabel (Ed.), *Handbook of research on science teaching and learning*. New York: MacMillan Publishing Co.

Usher, P. (1996). Feminist approaches to research. In D. Scott & R. Usher (Eds.), *Understanding educational research* (pp. 120-142). New York: Routledge.

Valero, P. & Zevenbergen, R. (Eds.) (2004). *Researching the socio-political dimensions of mathematics education: Issues of power in theory and methodology*. The Netherlands: Kluwer Academic Publishers Group.

von Glasersfeld, E. (1995). A constructivist approach to teaching. In L.P. Steffe & J. Gale (Ed.), *Constructivism in education* (pp. 3-15). New Jersey: Lawrence Erlbaum Associates, Inc.

von Glasersfeld, E., & Steffe, L.P. (1991). Conceptual models in educational research and practice. *Journal of Educational Thought, 25*, 91-103.

Wandersee, J. H., Mintzes, J.J., & Novak, J. D. (1994). Research on alternative conceptions in science. In D. Gabel (Ed.), *Handbook of research on science teaching and learning* (pp. 177-210). New York: MacMillan Publishing Co.

Webster's Third New International Dictionary of the English Language (Unabridged) (1976). Springfield, Mass: G & C Merriam Company, Publishers.

Weedon, C. (1987). *Feminist practice and poststructural theory*. Cambridge, Massachusetts: Blackwell Publishers

Winterdyk, J., Coates, L., & Brodie, S. (2006). *Qualitative and quantitative research methods reader: A Canadian orientation*. Toronto, ON: Pearson Education Canada Inc.

Wotherspoon, T. (1998). *The sociology of education in Canada: Critical perspectives*. Toronto: Oxford University Press.

Zitzewitz, P.W. & Neff, R.F. (1992). *Physics: Principles & Problems*. Ohio: Glencoe/McGraw-Hill Publishing Company.

Zwicky, J. (1995). Bringhurst's presocratics: Lyric and ecology. In T. Lilburn (Ed.), *Poetry and knowing* (pp. 65-117). Ontario: Quarry Press, Inc.

SUBJECT/OBJECT INDEX

NEW DIRECTIONS IN MATHEMATICS AND SCIENCE EDUCATION

Volume 1
Learning Science ***A Singular Plural Perspective***
Wolff-Micheal Roth ***University of Victoria, Canada***
ISBN 978-90-77874-26-4 (hardback) ISBN 978-90-77874-25-7 (paperback)
372 pages
Volume 2
Theorems In School ***From History, Epistemology and Cognition to Classroom Practice***
Paolo Boero (ed.) ***Universita di Genova, Italy***
ISBN 978-90-77874-22-6 (hardback) ISBN 978-90-77874-21-9 (paperback) 348 pages

Volume 3
The Culture Of Science Education *Its History in Person*
Kenneth Tobin *City University of New York* and **Wolff-Michael Roth (Eds.)** *University of Victoria, Canada*
ISBN 978-90-77874-35-6 (hardback) ISBN 978-90-77874-33-2 (paperback)

Volume 4
Teaching To Learn *A View from the field*
Kenneth Tobin *City University of New York and* **Wolff-Micheal Roth (eds.)**
University of Victoria, Canada
ISBN 978-90-77874-91-2 (hardback) ISBN 978-90-77874-81-3 (paperback) 284 pages

Volume 5
Understanding Teacher Expertise In Primary Science ***A Sociocultural Approach***
Anna Traianou ***Goldsmiths College, University of London, UK***
ISBN 978-90-77874-89-9 (hardback) ISBN 978-90-77874-88-2 (paperback) 178 pages

Volume 6
On The Outskirts Of Engineering *Learning Identity, Gender, and Power via Engineering Practice*
Karen L. Tonso *Wayne State University, USA*
ISBN 978-90-77874-96-7 (hardback) ISBN 978-90-77874-94-3 (paperback) 298 pages

Volume 7
Science, Learning, Identity *Sociocultural and Cultural-Historical Perspectives*
Wolff-Micheal Roth *University of Victoria, Canada* and **Kenneth Tobin (eds.)**
City University of New York
ISBN 978-90-8790-080-9 (hardback) ISBN 978-90-8790-090-8 (paperback) 364 pages

Volume 8
How Should I Know? Preservice Teachers' Images of ***Knowing*** (***by Heart***) in Mathematics and Science
Kathleen T. Nolan *University of Regina, USA*
ISBN 978-90-8790-213-1 (hardback) ISBN 978-90-8790-212-4 (paperback) 252 pages

Volume 9
Great Ideas in Science Education *Case Studies of Noted Living Science Educators*
Xiufeng Liu Nolan *University at Buffalo, SUNY, USA*
ISBN 978-90-8790-227-8 (hardback) ISBN 978-90-8790-226-1 (paperback) 124 pages

MAKING CONNECTIONS
Comparing Mathematics Classrooms Around The World

David Clarke ***University of Melbourne, Australia,***
Jonas Emanuelsson *Göteborgs Universitet, Sweden,*
Eva Jablonka *Freie Universität Berlin, Germany* and
Ida Ah Chee Mok (eds.) *The University of Hong Kong, Hong Kong SAR China*

In this book, comparisons are made between the practices of classrooms in a variety of different school systems around the world. The abiding challenge for classroom research is the realization of structure in diversity. The structure in this case takes the form of patterns of participation: regularities in the social practices of mathematics classrooms. The expansion of our field of view to include international rather than just local classrooms increases the diversity and heightens the challenge of the search for structure, while increasing the significance of any structures, once found. In particular, this book reports on the use of 'lesson events' as an entry point for the analysis of lesson structure. International research offers opportunities to study settings and characteristics untenable in the researcher's local situation. Importantly, international comparative studies can reveal possibilities for practice that would go unrecognized within the established norms of educational practice of one country or one culture. Our capacity to conceive of alternatives to our current practice is constrained by deep-rooted assumptions, reflecting cultural and societal values that we lack the perspective to question. The comparisons made possible by international research facilitate our identification and interrogation of these assumptions. Such interrogation opens up possibilities for innovation that might not otherwise be identified, expanding the repertoire of mathematics teachers internationally, and providing the basis for theory development.

THE LEARNER'S PERSPECTIVE STUDY Volume 2
ISBN 978-90-77874-90-5 hardback ISBN 978-90-77874-79-0 paperback
280 pages

IN SEARCH OF MEANING AND COHERENCE
A Life in Research

Wolff-Michael Roth
University of Victoria, Canada

In the course of his research career, much of which was based in his own classrooms, Wolff-Michael Roth explored numerous new theoretical frameworks when the old ones proved to be unable to account for the data. In this book, surrounding 11 of his publication s spanning 20 years of work, the author tells a story of how science education research concretely realized and singularized itself. That is, rather than taking sole credit for the work that ultimately came to bear his name, Roth develops a historical narrative in which his work came to realize cultural-historical possibilities inherent in the field of science education. But perhaps because some types of this work came to be realized for a first time, Roth's research also came to be characterized by others in the community as "cutting edge." This work, therefore presents as much an auto/biographical narrative as it presents a cultural-historical recollection of science education as it unfolded over the past two decades.

CULTURAL PERPECTIVES IN SCIENCE EDUCATION:
DISTINGUISHED CONTRIBUTORS
Volume 1
ISBN 978-90-8790-223-0 hardback ISBN 978-90-8790-224-7 paperback

KEY WORKS IN RADICAL CONSTRUCTIVISM
(edited by Marie Larochelle)

Ernst von Glasersfeld

Key works on radical constructivism brings together a number of essays by Ernst von Glasersfeld that illustrate the application of a radical constructivist way of thinking in the areas of education, language, theory of knowledge, and the analysis of a few concepts that are indispensable in almost everything we think and do. Ernst von Glasersfeld's work opens a window on how we know what we know. The present work grew out of a desire to make more accessible this line of thought, to highlight its originality and consistency, and to illustrate its fecundity in the domains of cognition and learning.

The first three parts of this book contain texts by Glasersfeld that outline the constructivist approach and explicate the frequently drastic reconceptualizations he has suggested. Both the last part and the postscript consist of commentaries by Edith Ackermann, Jacques Désautels, Gérard Fourez, Leslie P. Steffe and Kenneth Tobin, scholars in the fields that Glasersfeld has been concerned with. They examine a number of critical aspects pertaining to (radical) constructivism's current and future development, often tracing out paths that warrant further exploration and reflection, in particular concerning the sociopolitical dimension of knowledge. Key works on radical constructivism is intended as a reference book for researchers, educators, and students of education—and for anyone interested in grasping, or deepening their grasp of, radical constructivism's tenets, ambitions and concerns. Readers will discover in this collection of firsthand contributions the contours of a bold, contemporary debate about a most compelling current of thought.

Bold Visions in Educational Research volume 18
ISBN 978-90-8790-086-1 (hardback) ISBN 978-90-8790-085-4 (paperback)
340 pages

Printed in the United Kingdom
by Lightning Source UK Ltd.
125617UK00001B/91/A